Inter- and Intra-Vehicle Communications

OTHER TELECOMMUNICATIONS BOOKS FROM AUERBACH

Architecting the Telecommunication Evolution: Toward Converged Network Services
Vijay K. Gurbani and Xian-He Sun
ISBN: 0-8493-9567-4

Business Strategies for the Next-Generation Network
Nigel Seel
ISBN: 0-8493-8035-9

Chaos Applications in Telecommunications
Peter Stavroulakis
ISBN: 0-8493-3832-8

Context-Aware Pervasive Systems: Architectures for a New Breed of Applications
Seng Loke
ISBN: 0-8493-7255-0

Fundamentals of DSL Technology
Philip Golden, Herve Dedieu, Krista S Jacobsen
ISBN: 0-8493-1913-7

Introduction to Mobile Communications: Technology,, Services, Markets
Tony Wakefield, Dave McNally, David Bowler, Alan Mayne
ISBN: 1-4200-4653-5

IP Multimedia Subsystem: Service Infrastructure to Converge NGN, 3G and the Internet
Rebecca Copeland
ISBN: 0-8493-9250-0

MPLS for Metropolitan Area Networks
Nam-Kee Tan
ISBN: 0-8493-2212-X

Performance Modeling and Analysis of Bluetooth Networks: Polling, Scheduling, and Traffic Control
Jelena Misic and Vojislav B Misic
ISBN: 0-8493-3157-9

A Practical Guide to Content Delivery Networks
Gilbert Held
ISBN: 0-8493-3649-X

Resource, Mobility, and Security Management in Wireless Networks and Mobile Communications
Yan Zhang, Honglin Hu, and Masayuki Fujise
ISBN: 0-8493-8036-7

Security in Distributed, Grid, Mobile, and Pervasive Computing
Yang Xiao
ISBN: 0-8493-7921-0

TCP Performance over UMTS-HSDPA Systems
Mohamad Assaad and Djamal Zeghlache
ISBN: 0-8493-6838-3

Testing Integrated QoS of VoIP: Packets to Perceptual Voice Quality
Vlatko Lipovac
ISBN: 0-8493-3521-3

The Handbook of Mobile Middleware
Paolo Bellavista and Antonio Corradi
ISBN: 0-8493-3833-6

Traffic Management in IP-Based Communications
Trinh Anh Tuan
ISBN: 0-8493-9577-1

Understanding Broadband over Power Line
Gilbert Held
ISBN: 0-8493-9846-0

Understanding IPTV
Gilbert Held
ISBN: 0-8493-7415-4

WiMAX: A Wireless Technology Revolution
G.S.V. Radha Krishna Rao, G. Radhamani
ISBN: 0-8493-7059-0

WiMAX: Taking Wireless to the MAX
Deepak Pareek
ISBN: 0-8493-7186-4

Wireless Mesh Networking: Architectures, Protocols and Standards
Yan Zhang, Jijun Luo and Honglin Hu
ISBN: 0-8493-7399-9

Wireless Mesh Networks
Gilbert Held
ISBN: 0-8493-2960-4

AUERBACH PUBLICATIONS

www.auerbach-publications.com
To Order Call: 1-800-272-7737 • Fax: 1-800-374-3401
E-mail: orders@crcpress.com

Inter- and Intra-Vehicle Communications

Gilbert Held

CRC Press
Taylor & Francis Group
Boca Raton London New York

CRC Press is an imprint of the
Taylor & Francis Group, an **informa** business

AN AUERBACH BOOK

CRC Press
Taylor & Francis Group
6000 Broken Sound Parkway NW, Suite 300
Boca Raton, FL 33487-2742

First issued in paperback 2019

© 2008 by Taylor & Francis Group, LLC
Auerbach is an imprint of Taylor & Francis Group, an Informa business

No claim to original U.S. Government works

ISBN-13: 978-1-4200-5221-3 (hbk)
ISBN-13: 978-0-367-38831-7 (pbk)

Library of Congress Cataloging-in-Publication Data

Held, Gilbert, 1943-
    Inter- and intra-vehicle communications / Gilbert Held.
        p. cm.
    Includes bibliographical references and index.
    ISBN 978-1-4200-5221-3 (hardback : alk. paper)
    1. Motor vehicles--Automatic control. 2. Intelligent control systems. 3. Traffic engineering. 4. Transportation, Automotive--Communication systems. I. Title.

    TL240.H445 2007
    629.2'7--dc22   2007019850

Visit the Taylor & Francis Web site at
http://www.taylorandfrancis.com

and the Auerbach Web site at
http://www.auerbach-publications.com

Dedication

For over 20 years I have taught different graduate courses at Georgia College and State University. During this time I have been blessed to have students that livened up the classroom experience by not only being attentive, but also asking thought-provoking questions. In recognition of those students, this book is dedicated to those who asked both "why" and "how" as we examined various communications concepts.

Contents

Chapter 2

Communications Fundamentals 27

Chapter 3

Communications Technologies 51

Chapter 4

The Local Interconnect Network 79

Chapter 6

Intra-Vehicle Communications...113

Chapter 7

Inter-Vehicle Communications...137

Preface

Approximately every 5 or 10 years we are blessed with the implementation of a technology that can have a major bearing upon how we work, facilitates our productivity, and even enhances our recreational capability. In the past we witnessed the PC revolution, the advent of the PDA, and the growth in the use of wireless local area networks (LANs). Today a new technology, referred to as intra- and inter-vehicle communications, considerably influences how we operate our vehicles.

The use of communications within and between vehicles provides a mechanism to facilitate vehicle location, promote security, avoid accidents, and enhance our ability to arrive at our destination in a timely manner, avoiding traffic bottlenecks through the development of real-time communications flowing to in-vehicle navigation systems. Although only very basic information is now passed to vehicles, under the hood a revolution is occurring in the manner by which different modules communicate with one another that can be expected to be significantly built upon as other systems are added to provide additional communications-based functionality to vehicles planned for manufacture.

In the area of inter-vehicle communications several trials are occurring that could significantly reduce the number of accidents that occur each year. Although many readers are familiar with the use of an electronic pass that enables a vehicle to go through a toll booth without having to stop, the so-called E-ZPass electronic toll collection program uses a relatively short range tag the size of a deck of cards that is mounted on the window of a vehicle that an antenna at the toll booth reads. This type of communications is primarily unidirectional and only occurs as a vehicle comes within range of the antenna mounted at a toll both. As we will note later in this book, the ability to obtain an inter-vehicle communications capability is much more complex and requires the vehicle to become a member of a network consisting of other vehicles as well as lane markers, exit ramps, and other highway structures if we wish to obtain the ability to use the network to avoid both vehicle accidents and lane and exit drifting, which could also result in safety issues.

Because it is the job of an author to fully inform readers of all sides of an issue, we will note that there are some significant problems associated with intra- and inter-vehicle communications. Problems such as security represent issues that must be considered regardless of the size of a vehicle communications network. Still, other problems, such as radio frequency interference, can represent both controllable and noncontrollable issues, because it may be difficult or impossible to control the use of machinery and street lighting, let alone the periodic sunspots that radiate hundreds of millions of miles onto our small planet.

In addition, because vehicles can travel in areas ranging from an Alaskan highway to the Gobi Desert, communications within a vehicle must be able to operate under severe conditions, to include a significant temperature range and extreme vibrations. Another problem that adds to the temperature range is engine heat and the effect of the sun on a vehicle's outer skin surface. If you have a dark-colored vehicle used in Georgia during the summer, you probably thought about your ability to fry eggs on the hood without having to start the vehicle. Now imagine the effect of heat if you need to install wiring to remote sensors located on each corner of the vehicle as well as along its sides. Thus, at applicable locations in this book we will note the problems associated with intra- and inter-vehicle communications networking as well as actual and potential solutions to such problems.

As a professional author I truly welcome reader comments. Let me know if you feel I should expand upon a topic, if I provided too much information, and what topics you might like to read about in a future edition of this book. Of course, any other comments or suggestions are also welcomed. You can contact me through my publisher, whose address is on the cover of this book, or you can send an e-mail to gil_held@yahoo.com.

Gilbert Held
Macon, Georgia

Acknowledgments

The preparation of a book in many respects is similar to a sport in that it is a team effort. This book is certainly no exception, as it required the efforts of many persons to publish.

Once again I am indebted to Richard O'Hanley at CRC Press for green-lighting another idea and providing backing for this project. I would be remiss if I did not also thank the production staff at CRC Press for turning my manuscript into the book you are reading. Concerning the manuscript, it is with a great sense of pride that I thank my wife, Beverly Jane Held, for her efforts in turning my handwritten notes into a professionally typed manuscript. Beverly typed my first book on a 128-kB Macintosh many years ago. Although technology has certainly changed over the years, Beverly's typing skills continue to maintain a level of accuracy that is truly appreciated.

About the Author

Gilbert Held is an award-winning author and lecturer who specializes in the application of communications technology. Over the past 30 years, Gil has authored approximately 100 books and 300 articles focused on communications technology and personal computing. Although the number of books Gil has authored may appear to be quite high, that number includes second, third, and even fourth editions of several books that were researched and written over a long period of time.

In recognition of Gil's writing talents, he twice received the Karp Award for technical excellence in writing. Gil has also received awards from the American Publishers Institute and *Federal Week*. After earning a BS in Electrical Engineering from Widener University, Gil earned an MSEE degree from New York University and the MSTM and MBA degrees from American University. Presently Gil is the director of 4-Degree Consulting, a Macon, Georgia-based organization that specializes in the application of communications technology.

Chapter 1

Introducing the Smart Vehicle

Today we are on the verge of witnessing a revolution in automotive technology. Although many components and systems are still in the prototype stage of development, other components and systems are now offered as either a standard or an extra-cost option on numerous product lines. For example, automobile navigation systems have not only reached a level of mass marketing for use in vehicles, but in addition, the display area used for vehicle navigation is now integrated to provide images from real-time cameras mounted to display vehicle blind spots, permitting the driver to view the surrounding area without having to extend his or her head outside the window or become a contortionist when attempting to determine how close his or her vehicle is to a curb when parking.

In this introductory chapter we will primarily focus our attention upon differentiating between intra-vehicle and inter-vehicle communications, briefly examining the existing and evolving functionality associated with each type. In addition, as we discuss inter-vehicle communications we will note that this technology actually references a broad series of technologies, ranging in scope from global positioning system (GPS) positioning to vehicular networks consisting of other vehicles and sensors. Now that we know the plan for this chapter, let us turn our attention to existing and evolving vehicle communications systems.

1.1 Intra-Vehicle Communications

Intra-vehicle communications reference communications that occur within a vehicle. Such communications enable vehicle diagnostics where a technician can plug a tester into a port in the vehicle network and may be able to examine the operational state of various components of the vehicle as well as fluid levels and engine performance.

1.1.1 Communications Protocols

The foundation upon which intra-vehicle communications occurs is through the use of a communications protocol. That protocol represents a set of rules that govern the orderly transmission of data among various components in the vehicle. Later in this book we will examine in considerable detail the Local Interconnect Network (LIN) and the Controller Area Network (CAN), two of the more popular networks used to support intra-vehicle communications.

1.1.2 Additional Intra-Vehicle Communication Functions

In addition to providing a mechanism to move information between various vehicle components, the term *intra-vehicle communications* can be used to reference communications of stand-alone devices, such as a DVD player mounted in the rear of a vehicle that can be used to entertain youngsters of all ages on long trips, or a camera mounted at the rear of a vehicle that is wired to a display in the front console and that is provided to enable the driver to have a better view of the driveway when he or she backs out the vehicle or for use when parking.

1.1.3 Systems and Sensors

Within a modern vehicle a number of systems and sensors are used to provide different levels of functionality. Table 1.1 lists a few of the major systems and sensors that can be considered to represent the use of intra-vehicle communications. Prior to the 1990s most of the systems listed in Table 1.1, and any associated sensors, could be considered to represent autonomous systems because they were not integrated into a communications network for the access of performance or diagnostic information. Today, many vehicles have a degree of integrated diagnostics, with more expensive vehicles including between 50 and 100 microprocessors and 100 or more sensors that allow messages and commands to flow over a common intra-vehicle wiring network that enhances diagnostic testing, vehicle operations, and its safety.

Table 1.1 Major Vehicle Systems and Sensors

Air bags
Air conditioning and climate control
Braking system
Crash sensors
Data recorder
Engine control unit
Electronic stability control
Electronic steering
Infotainment system
Integrated starter generator
Lighting system
Power distribution and connectivity
Power train system
Seat belt sensors
Tire pressure monitoring system
Window and door systems

In this book we will devote separate chapters to examining the operation of LIM and CAN. Once this is accomplished, a third chapter will tie all intra-vehicle communications together, to include how such technologies as Bluetooth and OnStar and other electronic transmission systems are used within a vehicle, as well as such autonomous systems as E-ZPass, which was described in the Preface. Because some electronic systems can be considered to represent both intra- and inter-vehicle communications, this author will discuss those systems with respect to their primary use and then reference them in other portions of the book.

To obtain an appreciation of the operation and functionality of the vehicle systems and sensors listed in Table 1.1, let us briefly examine each.

1.1.3.1   Air Bag System

An air bag system consists of a number of air bags that can be located at various points within a vehicle. At a minimum, a driver-side air bag is typically mounted in the steering wheel of a vehicle, while a front-passenger-side air bag is commonly mounted in the dashboard above the location of the glove box. Other air bags may be mounted at the driver and passenger knee level, on each side of the vehicle, and at other locations in the vehicle. Typically, crash sensors mounted in the front bumper area of the vehicle are used to deploy both frontal air bags and knee-level air bags, if so equipped, in the event of a frontal crash. Due to the possibility that the deployment of a passenger-side air bag could kill or severely injure a child strapped into a car seat, many modern vehicles have sensors that will not allow the passenger-side air bag to deploy unless a sufficient weight resides on the passenger seat.

While front air bags are essentially standard on most vehicles produced today, side air bags, knee-level air bags, and "head curtain" air bags are either optional or standard on high-end vehicles. The side air bags are commonly deployed by sensors located behind door pillars, while side head curtain air bags are typically deployed when a vehicle is on the verge of a rollover. Thus, the sensors that control the deployment of side-mounted head curtain air bags measure stability, while the other sensors used to deploy air bags measure impact.

1.1.3.2   Air Conditioning and Climate Control System

The air conditioning and climate control system regulates the temperature and humidity within a vehicle. Although most vehicles have a single set of controls, some higher-end luxury vehicles include both driver and passenger thermostats that enable two temperatures to be maintained in different parts of the vehicle.

The key to the ability of the air conditioner to maintain a cooling capability is the pressure of the refrigerant used. Some high-end vehicles include a pressure sensor that will illuminate a warning indicator when the pressure falls below a pre-defined value, while other vehicles may provide a sensor that transmits a code to a diagnostic box in the vehicle.

1.1.3.3   Braking System

The modern automotive braking system represents the refinement of basic technology that has occurred over a period of approximately 100 years. Figure 1.1 provides a basic illustration of the components of a modern vehicle braking system.

The modern vehicle braking system consists of disk brakes in the front of the vehicle and either disk or drum brakes in the rear, with a system of tubes and hoses

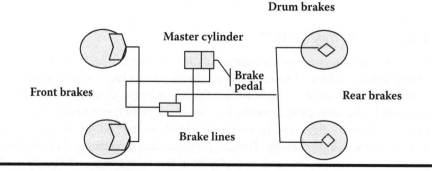

Figure 1.1   A vehicle brake system.

that connect the brake at each wheel to the master cylinder. When the brake pedal is depressed, it functions as a plunger in the master cylinder, which forces brake fluid through the brake lines to the braking unit at each wheel.

On a disk brake, the brake fluid from the master cylinder is forced into a caliper where it presses against a piston. The piston squeezes a pair of brake pads against a rotor that is attached to the wheel. This causes the wheel to slow down. When enough pressure is applied to the brake pedal, the resulting pressure causes the brake pads to force the wheel to stop.

Drum brakes are similar to disk brakes with respect to brake fluid being used to control their operation. Within a drum brake fluid is forced into the wheel cylinder, which pushes brake shoes forward so that their linings are pressed against the drum that is attached to the wheel. This action results in the wheel slowing and then stopping.

1.1.3.3.1   Master Cylinder

If you pop open the hood of your vehicle and look in the engine compartment in front of the driver's seat, you will notice a rectangular metal compartment with a top that either is pried off or has a screw cap you remove to add fluid or observe the fluid level. This metal compartment is the master cylinder. The master cylinder contains two separate master cylinders in a common housing, with two fluid wells that must be filled with brake fluid. Each fluid well operates a master cylinder that controls braking for two wheels, ensuring that if a brake line or brake should fail or the fluid level is insufficient in one well, only two brakes will become problematic and the operator should still be able to stop the vehicle.

1.1.3.3.2   Other Brake Systems

Other brake systems that are connected to the main vehicle brake system include the parking brake and, if offered, an antilock brake system.

The parking brake, which is also referred to as an emergency brake, controls the rear brakes through a series of steel cables. The steel cables are connected to either a foot pedal or hand lever and represent a mechanical system that bypasses the hydraulic system. In theory, this allows the vehicle to be stopped in the event of a total brake failure.

1.1.3.3.3   Antilock Braking System

The antilock braking system, more commonly referred to as ABS, is designed to prevent the wheels of a vehicle from locking while braking. This allows the vehicle operator to maintain steering control as well as shorten braking distance without skidding or losing control due to heavy braking.

An ABS consists of a central electronic unit, four speed sensors, each of which is connected to a wheel, and two or more hydraulic valves on the brake circuit. The electronic unit monitors the rotation speed of each wheel. The electronic unit is designed to sense that one or more wheels are rotating considerably more slowly than others, which would eventually cause locking. When this situation is noted, the control unit moves valves to decrease pressure on the braking mechanism, in effect reducing the braking force on the wheel. This results in the wheel turning faster. If the wheel or wheels turn too fast, the force is reapplied. This process of decreasing and increasing pressure results in a brake pedal pumping action. In more modern ABSs two additional sensors are used to support what is referred to as electronic stability. The two additional sensors are a wheel angle sensor and a gyroscopic sensor. When the gyroscopic sensor detects that the direction taken by the vehicle does not agree with what the wheel sensor indicates, the ABS software will brake the necessary wheel or wheels so that the automobile follows the direction of the driver. The wheel sensor can also be used to support cornering brake control because it informs the ABS that the wheels on the outside of a vehicle driving around a curve should brake more than the wheels on the inside, and by how much.

1.1.3.3.4   Brake Communications

Each of the braking systems previously described will convey some information to the vehicle operator in the form of illuminated indicators. For example, a low brake fluid level warning indicator will inform the vehicle operator that the master cylinder has a low level of brake fluid. However, by itself this indicator does not inform the operator if the low level of brake fluid resulted from gradual evaporation or from a hose break, faulty calipers, or another problem.

In actuality, the brake warning light indicator on most vehicles illuminates due to a pressure differential valve. That valve is usually mounted below the master cylinder and measures the pressure from each of the two sections of the master cylinder. If the pressure differential valve detects a difference, it indicates that there is more than likely a brake fluid leak in the system, although it does not pinpoint the leak. Similarly, if the ABS electronic control unit detects a malfunction in the system, it will illuminate an ABS warning light on the dashboard. This indicator will inform the driver of a problem in the ABS; however, it will not inform the driver of the cause of the problem. For both situations a drive to the automotive dealer or vehicle repair shop will be necessary to isolate the problem and initiate corrective action.

Although we live in an electronic age, there are some components that more often than not use a warning sensor that is not electronic. For example, some brake pads are manufactured with a "warning sensor" that will result in a squealing sound when the pads are worn to a point where they should be heard. The noise from the brake warning sensor is usually heard when the driver's foot is off the brake and stops when the driver presses his or her foot on the brake.

Instead of an audio warning, many higher-end vehicles use a more robust brake pad thickness sensor that will cause a warning light on the dash to illuminate when the thickness of the brake pad falls below a certain level.

1.1.3.4   Crash Sensors

The air bags installed in a vehicle are only as good as its control system. Most vehicles today use an electronic sensor system in which up to three crash sensors are installed in the forward crush zones of a vehicle. This enables the sensors to react almost instantly to the sudden deceleration that results from a frontal impact.

In addition to electronic sensors, there are switch trip or spring and mass sensors. The switch trip sensor includes a small metal roller in a housing that rolls forward when a sudden deceleration occurs. As the roller moves forward, it trips a switch that causes an air bag to deploy. The spring and mass crash sensors include a spring-loaded weight that is deflected by the occurrence of an impact. This action also results in the tripping of a switch that deploys an air bag.

1.1.3.4.1   Safety or Arming Sensor

To prevent a false deployment that could result from bumping into a curb or another object, or even a slow-speed fender bender, many air bag systems include a safety or arming sensor. These sensors are commonly located under the dashboard and passenger seat. The purpose of these additional sensors is to prohibit the deployment of an air bag unless it also experiences a rate of deceleration, which is usually less than that experienced by the crash sensors. Only when both crash sensors and one or more safety sensors are triggered by a collision will the electronic control module ignite the air bag inflator, which will cause the air bag to deploy.

Because a frontal crash commonly results in both driver and passenger air bags being deployed even when no passenger is in the vehicle, a pressure sensor is used in some vehicles to avoid this situation. That is, unless a weight is detected on the passenger seat, that air bag will not deploy in the event of a frontal impact.

1.1.3.4.2   Diagnostic Module

Modern air bag systems include a diagnostic module that is responsible for performing several key functions. Those functions include continuously monitoring the air bag system electrical circuits, controlling the operation of an air bag indicator lamp in the instrument panel on the dash, recording diagnostic information for use by a technician, and providing backup power in the event the vehicle's main battery power system is low or lost during a crash. Of the previously mentioned functions, only the control of the air bag indicator lamp is visible to the vehicle occupants. This indicator briefly illuminates when the engine is turned on as part of the monitoring process and will be solidly illuminated as a mechanism to alert the vehicle operator if a fault in the system is detected.

1.1.3.4.3   GPS Distress Signal

One of the more recent innovations in crash sensors is their use to trigger a GPS distress signal, which explains how GM's OnStar satellite system becomes informed about an accident. Other crash sensors will release electronically controlled door locks, close any open sunroof, and initiate the pretension of seat belts. Thus, there are a variety of crash sensors that support a variety of functions.

1.1.3.4.4   Data Recorder

Most vehicle manufacturers include an event data recorder (EDR), commonly referred to as a black box. This black box maintains the capability to record predefined vehicle system status information for a few seconds prior to a crash. In addition, vehicle speed, engine rpm, throttle position, and the brake switch ON/OFF status are recorded for a 5-s period preceding an air bag deployment or near-deployment situation. Unfortunately, vehicle manufacturers have not standardized the metrics they record, and there are slight to medium differences between vendors. Table 1.2 indicates the parameters recorded with many GM air bag systems.

To rectify the variable information contained in EDRs, the National Highway Transportation Safety Administration (NHTSA) will require vehicles starting with the 2011 model year to collect at least 15 types of data, including vehicle speed and whether the driver was wearing a seat belt. In addition, because many vehicle operators are unaware of the fact that their vehicle has an EDR, manufacturers will have to disclose the existence of the technology in owner manuals, also beginning with the 2011 model year. Unfortunately, the NHTSA rules only regulate what needs to be recorded and do not require that data recorders be included in all new vehicles.

1.1.3.5   Engine Control Unit

The engine control unit (ECU) is responsible for handling the logic necessary for managing a vehicle's power train in an efficient manner. Also referred to as the engine management system (EMS) and engine controller (EC), this device is a computer and memory that are part of the modern internal combustion engine. Through the use of an ECU actual engine performance can be monitored on a millisecond-by-millisecond basis. This enables the computer to be programmed to compensate the operation of the engine for such variables as ambient temperature,

Table 1.2 Parameters Recorded by GM Air Bag Systems

State of warning indicator when event occurred (ON/OFF)
Length of time the warning lamp was illuminated
Crash-sensing activation times or sensing criteria met
Time from vehicle impact to deployment
Diagnostic trouble codes present at time of event
Ignition cycle count at the time of the event
Maximum delta-V for near-deployment event
Delta-V versus time for frontal air bag deployment
Time from vehicle impact to time of maximum delta-V event
State of driver's seat belt switch
Time between near-deployment and deploy event (if within 5 s)
Passenger's air bag enabled or disabled state
Engine speed (5 s before impact)
Vehicle speed (5 s before impact)
Brake status (5 s before impact)
Throttle position (5 s before impact)

humidity, air density as a function of altitude, fuel octane rating, and driver operations, such as depressing the brake pedal. By monitoring and compensating for a large number of variables, the ECU can provide enhanced fuel efficiency and an increased level of power and responsiveness, while causing a lower level of pollution than engines without a modern ECU.

The main component of an ECU is a 16-bit or 32-bit microprocessor, such as IBM's PowerPC. The microprocessor must be fast enough to process engine sensor inputs in real time. The ECU contains the microprocessor and erasable programmable read-only memory (EPROM) or flash memory mounted on a printed circuit board to store software as firmware that is not erased if battery power is lost. Firmware can be reprogrammed, which enables advances in software development or the discovery of firmware errors to be corrected by uploading updated code.

1.1.3.5.1 Onboard Diagnostics

One of the key functions of the ECU is to facilitate the discovery of power train-related problems. To do so, modern ECUs support onboard diagnostics (OBD) that facilitate the diagnosis of engine faults, stored as fault codes in their internal memory. Thus, when a vehicle is serviced, a technician can connect a fault code reader or scanner to the engine and view any stored fault codes.

1.1.3.5.1.1 Fault Codes — When a fault occurs it triggers the malfunction indicator lamp (MIL), which is also referred to as a check engine light (CEL). The CEL can be a considerable annoyance, as it tells the vehicle operator that a problem exists; however, it does not indicate the seriousness of the problem. If no other warning lights are illuminated and the engine appears to be running normally, it is probably safe to travel home to have your preferred auto repair facility examine your vehicle. However, if you hear an unusual noise, encounter vibrations, or smell oil, you more than likely should stop immediately and call the nearest service station.

When the check engine light illuminates, a diagnostic fault code is recorded in the ECU's memory that corresponds to the fault. Fault codes fall into two categories: industry standard codes that are common to all vehicles in a class, and manufacturer-specific codes. Fault codes have evolved from the initial OBD, which primarily detected gross failures within circuits or sensors, to OBD II, which supports the detection of an engine misfire, fuel vapor leaks, if the catalytic converter is not performing correctly, and most emission problems that cause hydrocarbon emissions to exceed 1.5 times the federal limit.

Standard or generic fault codes are the same for all makes and models of vehicles and are required by law. In comparison, manufacturer-specific fault codes are unique to specific vehicles. Such codes primarily cover non-emission-related failures that occur beyond the engine control system, such as ABS codes, HVAC codes, air bag codes, and various electrical and body codes.

1.1.3.5.2 ECU Interface

Figure 1.2 illustrates in block diagram format the interfaces to an engine control unit. Note that the ECU may support a variety of communications. For example, it may support the Local Interconnect Network (LIN), the Controller Area Network (CAN), J1850, or J1959 to communicate with other modules. In addition, the ECU may support an RS232 or Ethernet connection for scanning fault codes in the system.

In examining Figure 1.2, note that communications between other modules and the ECU occur via the chassis connector. Thus, this connector would use a communications protocol to transmit and receive data, such as LIN or CAN.

Figure 1.2 Basic engine control unit connectors.

The sensor connector would support analog and digital inputs. In addition, the sensor connector can be connected to a lambda sensor that monitors the exhaust gas oxygen to balance the fuel mixture and a K-type thermocouple that is used to measure exhaust gas temperature.

1.1.3.6   Electronic Stability Control

Electronic stability control (ESC) can be considered to represent an enhancement as well as an evolution of the antilock brake system. Under ESC, sensors are used to monitor everything from the position of the steering wheel and tire speed to the centrifugal forces a vehicle experiences when cornering. If the ESC system detects a potential driver loss of control, it will automatically apply individual front or rear brakes and may also reduce excess power as needed to help correct understeering or oversteering. In an understeering situation, the front end of a vehicle tends to slide out or plow ahead. ESC will automatically apply the inside rear brake to assist the driver's turning and may also reduce engine power. If an oversteer situation occurs where the rear end of the vehicle tends to slide out or fishtail, ESC will automatically apply the outside front brake to assist the driver to make the turn and correct the fishtailing. Although ESC primarily assists drivers in road cornering, it also provides assistance in accelerating and braking, responding when it senses impending wheel lockup, wheel spin, or loss of vehicle control. Thus, ESC helps improve traction, maneuverability, and stability.

ESC dates to 1997, when it appeared on select Mercedes, Cadillacs, and the Chevrolet Corvette. Vehicle manufacturers offer ESC systems under different marketing names, such Vehicle Stability Assist™ (Acura) and Advance Trac® (Ford), while General Motors refers to it as StabiliTrak®. In spite of being available for 10 years, only 6 percent of vehicles are purchased with ESC technology.

1.1.3.7   Steering

Today the vast majority of vehicles on the road use a hydraulic steering system. When power steering made its commercial debut in 1951, the cost of a gallon of gas was around a quarter. A hydraulic steering mechanism that applied torque based upon the force applied to the steering wheel resulted in a pump directly connected to the engine. Although this consumed energy even when no steering was requested, the cost of gasoline was so low that few people thought a replacement was necessary.

Over the years the cost of gasoline considerably increased, resulting in engineers developing electronic-based power steering systems. In 1993 Honda introduced the first all-electric power steering system on its Acura NSX sports car. The Honda system used a brush DC motor installed concentrically around the rack that combined torque and velocity information from the steering wheel with vehicle speed data to compute the optimum amount of steering assistance necessary.

By the mid-1990s Honda migrated its EPS design to its more reasonably priced S2000 sports car. Because the DC motor has some unattractive features, such as brush arcing and brush friction, its performance is limited at higher motor speeds and can be awkward for supporting larger vehicles. Thus, engineers turned to several alternatives, to include the switched reluctance motor, which contains no permanent magnets and operates well in high-temperature environments, and is able to survive a single-phase failure and keep on operating; a brushless permanent magnet (BPM) motor that uses rare earth magnets for higher efficiency; and the AC induction motor (ACIM), which can achieve lower levels of torque than even BPM motors.

Since 2000 over 10 million EPS units have been manufactured, with the majority based upon the use of a BPM motor. Today all hybrid vehicles use electronic steering systems where an electric motor is used in place of a hydraulic pump. Although not a hybrid, the 2007 Mini Cooper provides a 20 percent improvement in fuel consumption in comparison to the 2006 model, with part of the credit due to its electronic steering, along with water and oil pumps that are designed to deliver only as much pressure as needed. As the price of gasoline continues to remain high, we can expect an increase in the use of electronic steering on other vehicles.

1.1.3.8   Infotainment System

Infotainment represents a relatively new word that means a combination of information and entertainment. By 2006 many luxury vehicles began to feature large liquid-crystal displays (LCDs) that are mounted in the dash or pop up from the dash when a button is pressed. These LCD screens are used for displaying navigation maps and directions, climate control, trip information, and even tuning the radio. Although all of these functions are important, until recently no vehicle manufacturer had integrated entertainment into its information system for all car models. The exception is vehicles like the Audi A8, BMW 5, 6, and 7 series, E and S classes of Mercedes-Benz, and Rolls-Royce and Maybach luxury vehicles. Such vehicles employ a Media-Oriented System Transport (MOST) high-speed digital networking standard for moving multimedia content throughout a vehicle. Serving as a backbone for in-vehicle infotainment systems, MOST allows car manufacturers and suppliers to interconnect AM/FM radios, TVs, DVD players, navigation systems, cell phones, PDAs, and in-vehicle computers as modular devices in an automotive environment. Because multimedia content on the MOST bus can be selected in all available combinations, it becomes possible for passengers to decide on which services they desire to use. For example, one passenger could listen to a CD via headphones, while a second passenger might watch a DVD on a TV mounted on the rear of the driver's seat. Although MOST technology is used in upscale vehicles, until now it has been too expensive for low-end vehicles. This is about to change, as Microsoft is currently adapting its Windows CE operating system to merge information and entertainment into an infotainment system.

1.1.3.8.1   Microsoft's Efforts

According to the Microsoft Windows Automotive Web site, the release of Windows Automotive "Version 4.2 marks the introduction of the first Microsoft .NET-connected, voice-enabled software for building next-generation, in-vehicle devices." This technology will enable the industry to give drivers and passengers seamless access to a wide range of Web services, as well as smooth functionality among all Windows-powered devices. Microsoft is shortly expected to offer its navigation system with an in-vehicle computer referred to as T Box. T Box is designed to support connectivity from portable devices, to include cell phones, laptops, PDAs, and MP3 players. Consisting of a processor, memory, and storage, T Box is expected to reach the North American market between 2007 and 2008.

The first vehicle with a prototype of the Microsoft T Box was a 2006 Alfa Romeo 159. Fiat Auto plans to use the Microsoft infotainment system in four upcoming Alfa Romeo and Fiat models, to include all Fiat Lancia and Alfa Romeo 159, Brera, and Spider vehicles. In demonstrating the Microsoft infotainment system, a mobile phone uses Bluetooth wireless technology to link to the T Box. When a call is received, the system both identifies the caller and automatically lowers the volume of any music being played to facilitate the conversation. Concerning music, the Microsoft system allows music to be played from any device that can be plugged into a T Box USB port. The T Box also connects to Apple Computer's iPod to play tracks that are not encoded in its proprietary advanced radio coding (AAC) format.

1.1.3.8.2   Other Developments

Although Microsoft may be the best known company, it is not the only company developing infotainment systems for vehicles. Other vendors actively working in this area include SMSC, Fujitsu Microelectronics, Chrysler, Mercedes, and General Motors, to name but a few.

Both SMSC and Fujitsu are designing MOST-compatible products. The MOST network is used for the distribution of multimedia content enabling in-vehicle consumer devices such as DVD players, MP3 players, GPSs, car phones, and Bluetooth devices and similar products to work as if they are a single system instead of acting as single independent devices, each requiring its own control. MOST both describes the physical connection between devices and defines the signals they exchange for interoperability.

Another trial that warrants a brief description is the efforts of Hyundai Autonet of Korea. Hyundai recently developed software modules that enable MPEG transmission over MOST. The goal of Hyundai is to exploit MOST technology for both audio and video transmission.

The first Hyundai module is a driver information system (DIS). The DIS incorporates MOST, a wireless network, and CAN for the distribution of multimedia content by MOST, while code division multiple access (CDMA) or Global System for Mobile (GSM) supports cell phones, and CAN is used for body control. The second module developed by Hyundai, referred to as a multimedia information system (MIS), targets rear-seat entertainment. This module uses the MOST network for the distribution of multimedia content and is focused on supporting multiple screens in a vehicle. Both the original equipment manufacturer (OEM) and the aftermarket are targeted for sales by Hyundai.

1.1.3.9   Integrated Starter Generator

To obtain an appreciation for the functionality of the integrated starter generation, let us review the basic vehicle starting system. The starting system uses the battery as its power source. When a key is inserted into the ignition switch and turned to the start position, a series of actions begin. First, a small amount of current is passed through a neutral safety switch to a starter solenoid or starter relay. The solenoid allows a high current to flow from the battery through the battery cables to the starter motor. The starter motor then cranks the engine, which enables the pistons in the engine to move downward, creating a suction that will draw a fuel–air mixture into the cylinders. Then a spark created by the ignition system ignites the mixture, enabling the engine to start. Figure 1.3 illustrates the basic vehicle starting system.

1.1.3.9.1   Generator

Because a battery can only hold so much power it needs to be recharged. Initially generators were used in vehicles. Generators operate in a manner similar to that of

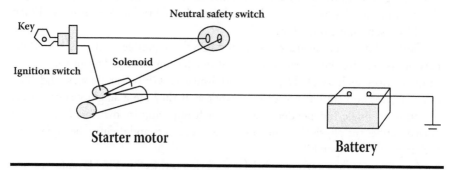

Figure 1.3   The basic vehicle starting system.

an electric motor, but in reverse. That is, instead of applying electricity to make it spin, a generator spins a series of windings of fine wire, referred to as the armature, by connecting the generator to a belt and pulley operated by the engine. As the armature is spun by the rotation of the belt and pulley, it generates a current and voltage that is proportional to the speed at which the armature spins.

The current produced by the generator is governed by the sate of the battery. If the battery is fully charged and no subsystems are using power, the current output is zero. The current output of the generator is controlled by the magnetic field based on its wiring, as well as the speed at which the armature moves. This results in a key shortcoming concerning the use of generators, because a generator-equipped vehicle cannot charge the battery when the vehicle idles, and resulted in the development of the alternator. Today, all modern vehicles use an alternator, so let us turn our attention to its operation.

1.1.3.9.2 Alternator

An alternator spins a magnetic field inside of windings of a wire referred to as a stator to generate electricity. Doing so allows the wiring from the alternator to carry a relatively high output current. Because the heavy windings are fixed instead of rotating, as in a generator, the alternator can be spun at higher speeds. This enables the alternator to reach its maximum output faster, as well as to be spun at engine idle speeds that enable a vehicle to produce a sufficient amount of electricity to power most, if not all, of the electrical requirements of a vehicle without having to drain the battery.

The output of the alternator stator normally consists of three separate windings of wire. These windings are set to generate ACs slightly out of phase with one another to smooth the electrical output. Inside the alternator a diode is used to convert or rectify the generated AC into DC. By arranging diodes so that current from each of three stator wires passes in one direction, and then summing the output of the diodes, a smooth and stable DC output is obtained that does not require any moving parts.

1.1.3.9.3 Regulators

There are two types of regulators: those that work with all generators and externally regulated alternators and those that work with internally regulated alternators. The first type represents a mechanical device that requires periodic adjustment, while the second type represents a more modern solid-state device. Both types of regulators create an average voltage and limit the amount of current.

If the alternator generates less voltage than the battery has in it, the indicator light on the dash will illuminate to indicate a problem. Because the light is connected to one side of the field current system inside the alternator and to a switched ignition power source on the other side, when you insert and turn the ignition key, the field acts as a ground and power flows through the light to ground, illuminating the indicator light for a short period. As the vehicle starts, the voltage at the field that is powered by the alternator will shortly match the battery voltage. When this occurs, the voltage on each side of the indicator light balances the other and the light goes off, indicating that all is well with your alternator. However, if the output of the alternator drops due to a belt problem or an electrical fault in the alternator, the voltage becomes unbalanced. Then, there will be less voltage on the field side of the indicator light, which results in some electricity flowing through the light, causing it to illuminate and indicate a problem has occurred. Now that we have an appreciation for the basic vehicle starting system, let us focus our attention upon the integrated starter generator (ISG).

1.1.3.9.4   Integrated Starter Generator and Alternator

The driving force behind the development of the integrated starter alternator (ISA) is the cost of fuel, which made manufacturers look for new methods to reduce full consumption. One method is a camless engine, in which valves are actuated by exhaust gas pressure instead of hydraulics or solenoids. Because a camless engine requires 42 V, the conventional use of an alternator and voltage regulator, which is limited to allowing voltages between 13 and 15 V, could not be used. Fortunately, the ISG and ISA permit between 40 and 600 V, enabling support for camless engines that can substantially reduce fuel consumption.

The first belt-driven ISG was developed in 2001. When functioning as a conventional alternator, the ISG generates electrical power when a vehicle is running, charges the battery if needed, as well as supplies electrical power to electrical devices in the vehicle. In 2003 the first ISA was developed and is now used in the Mercedes Maybach, an ultra-luxury vehicle.

Both the ISG and ISA are expected to enable a move to 42-V electrical architectures that can enhance fuel efficiency. The use of camless technology, as well as a start–stop capability that quietly shuts down the engine while vehicles wait at traffic lights and then quickly restarts the vehicle when the light changes, substantially improves fuel economy in urban areas.

1.1.3.9.4.1   Components

There are five major components used to create an ISG system. Those components include a three-phase AC motor integrated into the internal combustion motor, an AC/DC converter, a DC/DC converter, electronics for driving the ISG system, and an energy management system that controls the ISG. The AC/DC converter is bidirectional and rectifies the AC generated by a three-phase motor, while the DC/DC converter usually provides 12 V of DC or even 5 V of DC and 110/120 V of

AC in addition to 42 V of DC. Both the ISG and ISA can be viewed as the building blocks for future enhanced vehicle fuel economy.

1.1.3.10 Lighting System

The vehicle lighting system has a core set of lights, to include headlights, taillights, stop lights, reverse lights, turn signal indicators, and a variety of interior lights, ranging from those that illuminate when a portion of a visor is opened to reveal a mirror to map lights that illuminate when an interior button above the driver or passenger seat is pressed. In addition, depending upon the styling of the vehicle, lights may be mounted under the outside rearview mirrors and other areas to provide illumination at night when passenger doors are opened. In addition, fog lights may be mounted in the front of the vehicle to increase illumination of the road when driving at low speed or in poor visibility conditions.

1.1.3.10.1 Light Sources

There are three primary light sources used in vehicles: incandescent light bulbs, light-emitting diodes (LEDs), and neon tubes. The incandescent light bulb is commonly used for brake, turning, and reversing lights, tail lamps, parking lamps, and side turn signal repeaters. LEDs, due to their lower power consumption and longer life, are increasingly being used in automotive signaling lamps and are now commonly used in brake lamps and signaling lamps, as well as to satisfy many other light sources.

LEDs were first used for the center high-mounted stop lamp (CHMSL), which represents a central brake lamp mounted higher than a vehicle's left and high brake lamps. Neon tubes, long popular with the custom aftermarket as a mechanism to customize a vehicle's visibility at night, are also used in some vehicle CHMSLs. Although representing an incandescent light bulb, since the 1990s automobile manufacturers have used halide lamps for headlamps because these lamps produce approximately twice as much light as tungsten headlamps. Regardless of the light source, electricity needs to flow to light bulbs, LEDs, and neon tubes. This flow of electricity occurs via the lighting wire harness used in vehicles.

1.1.3.11 Power Train

The power train of a vehicle represents a key series of interconnected components that generate power that enables vehicle movement. Those components include the engine, transmission, driveshaft, differentials, and drive wheels. Vehicle manufacturers provide a long warranty on the power train, with some warranties up to 100,000 miles or 10 years. However, not all power trains are defined uniformly,

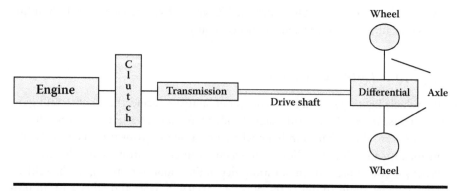

Figure 1.4   Single-axle power train.

with some warranties limited to covering the engine, transmission, and other components integral to the transmission.

Figure 1.4 illustrates in block diagram format the power train of a single-axle vehicle.

The power train is also commonly referred to as the drivetrain, as it moves or drives the vehicle. In examining Figure 1.4, note that the axle can be a pin, bar, or shaft on which a pair of wheels rotate. On the axle a set of gears, referred to as the differential, transfer the motion of the drive shaft turning to the axle, enabling the wheels to move. The engine develops rotational power that is transferred via the clutch to the transmission, which turns the drive shaft.

Within a modern vehicle there are a number of power train sensors that either communicate with lights on the dash or transmit one or more codes to the power train control module (PCM) when operations deviate from the norm. Thus, let us briefly discuss the PCM.

1.1.3.11.1   Power Train Control Module

The PCM can be considered to represent the key location to initially examine when power train problems occur. The PCM is an onboard computer that examines the operational state of the charging system, engine, transmission, and various emission controls and communicates with other onboard control modules. The fault codes stored are typically used as an initial guide for troubleshooting a vehicle problem. On many vehicles the PCM is in effect the previously described ECM.

1.1.3.12   Seat Belt Sensors

Due to vehicle safety regulations, the modern vehicle has a number of seat belt sensors that both cause a dash indicator to illuminate and generate an alarm when the engine is operating but the front seat belts are not connected. Because a passenger may not be present, a weight sensor is also used on the passenger side of the front seat. If the seat belt is not fastened and the weight sensor does not activate, then the indicator will not illuminate, nor will the alarm be generated.

1.1.3.13   Tire Pressure Monitoring System

Due to the recall of millions of tires during the past five years, several manufacturers introduced tire pressure monitoring systems as options, or as a standard with high-end luxury vehicles. In 2003 the U.S. Congress passed legislation that will eventually require all new vehicles to be equipped with a tire pressure monitoring system. Because maintaining correct tire pressure determines the load a vehicle can carry, too little pressure can cause an accident to occur. Thus, the tire pressure sensor system represents a safety system.

1.1.3.13.1   Types

There are two major types of tire pressure systems: direct and indirect. A direct system involves attaching a pressure sensor transmitter to the vehicle's wheel inside the tire's air chamber. An in-vehicle receiver then warns the driver if the pressure in one tire falls below a preset threshold. In comparison, an indirect system uses the vehicle's antilock braking system's wheel speed sensors to compare the rotational speed of one tire versus that of the other tires. If a tire has low pressure, the tire will then rotate at a different number of rotations per mile than the other tires. This fact then enables the problem to be brought to the attention of the driver.

1.1.3.14   Window and Door System

The modern vehicle has a series of window and door sensors and controls that effect warning lights, interior and exterior lights, and the ability to operate windows other than the driver's window and doors other than the driver's door. For example, opening a vehicle door at night will cause the dome light to illuminate, and if the vehicle is equipped with lights mounted under the rearview mirror or similar lighting, those lights will illuminate. If the vehicle's engine is on and a door is opened, a warning light and audible tone alert the driver.

The driver can usually control the ability of passengers to operate doors and windows by a series of controls on the driver-side door panel. If the door lock is disabled and a passenger opens a door, this action will either result in the illumination of an indicator, such as "door ajar," or display the door on a graphic image of the vehicle if the vehicle has such a display. Thus, what we normally take for granted when opening and closing windows is actually a complex series of sensors, wiring, and control indicators.

1.2 Inter-Vehicle Communications

In comparison to intra-vehicle communications, which is used to describe communications within a vehicle, the term *inter-vehicle communications* represents communications between vehicles or vehicles and sensors placed in or on various locations, such as roadways, signs, parking areas, and even the home garage. Inter-vehicle communications can be considered to be more technically challenging because vehicle communications need to be supported both when vehicles are stationary and when they are moving.

Historically, both vehicle and roadside communications systems were autonomous. For example, vehicle-mounted cameras display images on the vehicle console, while roadside communications at one time were limited to drivers observing road signs informing them to turn their radio dial to a certain frequency to obtain traffic information. Perhaps the earliest example of inter-vehicle communications is the use of a prepaid or automatic billing system when a vehicle slows but does not stop at a toll booth, using a small electronic transmitter to communicate with a receiver connected to an antenna at the toll booth. Another early example of inter-vehicle communications that many readers might wish to forget is the integration of cameras and speed sensors that determine the speed of a vehicle and, if over the speed limit, take a picture of the license plate, which is used to create a speeding ticket that is then mailed to the driver's address.

The previously mentioned examples of inter-vehicle communications represent legacy applications that will continue to exist for the foreseeable future. More modern inter-vehicle communications will provide a range of applications. Although some of these applications are presently in field trials, other applications are now operational on certain vehicles, and a few applications are simply being considered for future development. Table 1.3 lists three major categories of inter-vehicle communications applications that either are available on certain vehicles or can be expected to become available in the near future.

In examining the entries in Table 1.3, it is apparent that inter-vehicle communications is focused upon safety. As we probe deeper into each of the applications listed Table 1.3, we will note that there are various subapplications included. The degree of incorporation of such subapplications into an application can vary by vehicle and vehicle manufacturer due to the lack of standards defining inter-vehicle communications at the present time. In the following sections we will discuss each application listed in Table 1.3 and many of the subapplications.

Table 1.3   Major Inter-Vehicle Communications Applications

Cooperative driving
Consumer assistance
Smart parking

1.2.1   Cooperative Driving

Cooperative driving involves the use of onboard sensors in a vehicle that communicates with sensors on the road, mounted on signs, and on other vehicles to assist drivers in safely moving their vehicle from point A to point B. Table 1.4 lists some of the subapplications that can be incorporated into a cooperative driving application.

The ability to integrate cooperative driving applications into a vehicle depends upon several factors. Those factors include sensors located on other vehicles, implanted in the roadway, and mounted on signs, a communications capability that enables messages to be exchanged with other vehicles and receive data from sensors, as well as a type of radar that can be used to observe other vehicles not equipped with sensors, and terrain features and potential obstacles as a vehicle moves over a roadway. In addition, an ability to form an ad hoc network with other vehicles equipped with sensors can be considered to represent a future goal that will allow all vehicles to observe the status of other vehicles and take corrective action when necessary. To support cooperative driving applications, one or more onboard computers will be required to process signals received from sensors in the roadway, mounted on signs, and in other vehicles. Such microprocessors will also have to control onboard vehicle communications to include transmitting short-range radar or sonar signals and receiving and interpreting reflections, as well as using received data to assist the driver in controlling the vehicle. Thus, some aspects of inter-vehicle communications can be considered to represent a Buck Rogers future; however, instead of being science fiction, many vehicle manufacturers as well as third-party researchers are actively developing the components that will make inter-vehicle communications a reality. Now that we have an appreciation for the general manner by which cooperative driving applications can occur, let us turn our attention to some specific subapplications.

Table 1.4   Cooperative Driving Subapplications: Accident Warning

Frontal collision prevention
Hazard warning
Intersection alert
Overtaking and lane change assistance
Rear-end collision prevention
Road departure prevention
Speed alert

1.2.1.1   Accident Warning

When you travel on an interstate highway or another type of major roadway, you may periodically observe highway signs that convey different information. For example, a sign might be used to display an AMBER Alert when a child is missing. That same sign may also be used to indicate the occurrence of an accident and its effect upon travel delays, or even suggest an alternate path the vehicle driver should consider. This type of accident warning is passive in that it does not force a driver to do anything and, in addition, could be overlooked if the driver was changing lanes or performing another vehicle operation that results in him or her not observing the contents of the sign. In comparison, an active accident warning system conveys information to the driver that is hard to ignore. For example, radio transmitters could broadcast an accident warning message that could be displayed on the vehicle dashboard. In addition, an onboard vehicle radar could be used to sense a traffic buildup and automatically slow the vehicle. In other accident warning systems sensors in vehicles could determine that a crash occurred if air bags were deployed and relay this information via an ad hoc communications network to other vehicles on the roadway as well as inform OnStar or a similar service. Later in this book, when we focus on inter-vehicle communications in more detail, we will examine how ad hoc networks operate.

1.2.1.2   Frontal Collision Prevention

Through the use of radar or another type of electronic or light pulse, it becomes possible to measure the round-trip delay when pulses bounce off an object. The delay decreasing indicates that the distance between the vehicle and another vehicle ahead of it or any obstruction in the roadway is decreasing. This information can be used to automatically adjust the speed of the vehicle and inform the driver via a visual or audio signal or alarm of the potential occurrence of a frontal collision condition.

1.2.1.3   Hazard Warning

Similar to the previously described frontal collision prevention, a hazard warning can occur via measuring the round-trip delay associated with transmitting pulses from all four sides of a vehicle. However, due to cost as well as possible interference issues, it may be more practical to use sensors that operate on other frequencies, such as wireless local area networks (LANs), to communicate the location of other vehicles and even the center of the roadway to moving vehicles. In the next two chapters in this book we will review the basic use of communications technologies for vehicles, to include the fundamental concepts of wireless communications.

1.2.1.4 Intersection Alert

Any analysis of collisions will indicate that a substantial number of vehicle accidents occur at intersections. Those intersections can include two-way and four-way stops and traffic lights. Unfortunately, some persons forget to come to a full stop at a stop sign, while other persons run yellow and red traffic lights.

One method to reduce traffic accidents at intersections is to warn drivers as they approach the intersection. This warning or alert can result from an active or passive action. An active action could result from a transmitter sending a low-powered message or code at predefined intervals that an automobile receiver acknowledges by transmitting either an audio or video alert to the driver. In a passive environment a reflector coating could be used that, when pulsed by an onboard vehicle system, would result in an intersection alert. Another method for warning a driver could occur through the use of the built-in navigation system, which could produce an audio alert that one is approaching an intersection.

Another method to reduce the danger at intersections can be obtained through the use of an ad hoc network whereby vehicles reaching a network inform other vehicles of traffic flowing toward the intersection. Unfortunately, this technique depends upon vehicles approaching an intersection being equipped with a pulse system as well as the ability to participate as a node in an ad hoc network. Probably a more practical method is to place sensors at intersections that broadcast on a specific frequency the occurrence of two or more vehicles approaching the intersection, enabling other vehicles tuned to that frequency to note that they are approaching a busy traffic intersection.

1.2.1.5 Overtaking and Lane Change Assistance

Overtaking or passing a vehicle can be dangerous due to a blind spot in the driver's rearview mirror. Thus, an onboard system consisting of a side-mounted camera or the use of directional pulses that time reflections to indicate the presence or absence of a vehicle can facilitate a driver passing another vehicle safely as well as changing lanes.

1.2.1.6 Rear-End Collision Prevention

Previously we discussed front-end collision prevention. Rear-end collision protection can be reduced by other vehicles having a front-end collision system. Because it may take decades for a majority of vehicles to have such a system, another method is required to warn vehicles of a potential rear-end collision. That method can include the use of a camera that provides vehicle drivers with better visibility to the rear of their vehicle or the use of radar-equivalent pulses to the rear of the vehicle. The latter can use the reflection time as a mechanism to indicate to the vehicle operator a pending rear-end collision or potential danger due to a vehicle closing the distance between vehicles. Similarly, the following vehicle would use a sensor to slow down the speed of the vehicle as it approaches another vehicle in the same lane.

1.2.1.7   Road Departure Prevention

There are two basic methods that can be employed to alert a vehicle operators that their vehicle is moving off the roadway. The first method involves sensors placed on the center or edge of the roadway. As a vehicle moves away from the center sensors or crosses an imaginary line formed by the edge sensors, audio or video alarms could be generated and perhaps also steering adjusted. The second method involves using a microprocessor to observe steering action. When a person becomes tired or falls asleep at the wheel, the minimal steering of a vehicle on a roadway either stops or becomes awkward, allowing the condition to be recognized and the driver alerted.

1.2.1.8   Speed Alert

There are several methods that can be used to alert the vehicle operator to a speeding condition. First, a predefined threshold can be either set or sensed from the roadway. Then, if the vehicle exceeds either limit the driver would be alerted. A second method could involve automatically slowing the vehicle after sensors observe that the vehicle successfully passed another automobile and returned to its original lane.

1.2.2   Consumer Assistance

A second emerging area associated with inter-vehicle communications is consumer assistance. Although most high-end vehicles have built-in navigation systems that provide route planning and denote local points of interest, there are other functions that are emerging to expand the field of consumer assistance. Those additional functions include traffic information, mobile business support, and multimedia services.

1.2.2.1   Traffic Information

Until recently, the majority of traffic information was obtained by drivers either observing the contents of a highway sign or turning their radio to a specific frequency based upon the contents of a highway display. Within the past few years several vendors have offered subscribers real-time traffic information delivered to either their smart phone or in-dash navigation system. Recently, one vehicle manufacturer initiated an agreement with a third party to provide a year of free traffic information to persons purchasing certain types of vehicles. If this catches on, we can reasonably expect other vehicle manufacturers to provide a similar service to purchasers of their vehicles.

1.2.2.2   Mobile Business Support

Although the cell phone is as ubiquitous as the personal computer for the vast majority of persons in a vehicle, it functions as a stand-alone device. In the future we can expect the integration of cell phones to provide mobile business support, allowing drivers and passengers to check inventory and delivery dates, examine documents that are displayed on a vehicle's navigation system, and perform other business-related tasks. With the integration of the cell phone to the larger screen in a vehicle's navigation system, it becomes possible for a person to check different computer system applications, e-mail, and perform Web surfing without the constraints associated with the narrow cell phone display.

1.2.2.3   Multimedia Services

In the first section of this chapter we discussed infotainment as being based upon intra-vehicle communications. Although the display of a DVD on screens built into the rear of vehicles is certainly an example of an intra-vehicle communications application, it is also possible to receive multimedia services from outside the vehicle. When this occurs, communications can be considered to be inter-vehicle.

1.2.3   Smart Parking

There are two types of smart parking: one type can be categorized as intra-vehicle communications, while the second type represents inter-vehicle communications. The first type of smart parking involves the use of cameras and sensors to park the vehicle automatically. Referred to as a parking assistant system, this feature is now available on the 2007 Lexus LS 460. The second type of smart parking involves having parking spots equipped with wireless sensors, while wireless guidance devices are embedded into vehicles. This allows both vehicle-to-vehicle and vehicle-to-sensor communications that facilitate a vehicle automatically locating and parking, while noting the available parking space may include other vehicles parked, so that they overlap a portion of the available space.

1.3 Summary

Inter-vehicle communications is a wide-ranging field. It involves the use of sensors and wireless systems that use the Doppler effect to judge the distance to certain objects, to include other vehicles and roadway signs and sensors. In addition, some applications require the use of a wireless ad hoc network that enables vehicles to function as network nodes and relay information to other vehicles. Later in this book we will examine in considerable detail ad hoc networking and how this technology can be implemented to support emerging inter-vehicle communications. For now, we can simply note that the vehicle of tomorrow will be considerably different from the vehicle of today due to the use of short distance transmitters and receivers that operate similarly to Doppler radar, an increase in a vehicle's ability to recognize many type of sensors and take some action, as well as the ability of vehicles to become part of an ad hoc network to facilitate communications so that when one vehicle learns something, it can rapidly share its knowledge with other vehicles.

Chapter 2

Communications Fundamentals

It is a given that the readers of this book will have a diverse background with respect to communications concepts. Thus, to make the contents of this book as meaningful as possible to a diverse readership, this chapter will review a series of communications fundamentals. Those fundamental communications concepts we will cover include bandwidth, frequency, and noise that affect both wired and wireless communications. Once this is accomplished, we will examine the allocation of the radio frequency spectrum, noting how the frequency spectrum is allocated in the United States, the nomenclature used to categorize different segments or bands of frequency, and 18 common and evolving wireless applications, to include where in the frequency spectrum inter-vehicle communications is expected to occur. Once this is accomplished we will then examine radar operations that can be used to denote the distance to objects, as their use can provide the ability of one vehicle to note the distance of other vehicles and potential road obstructions. Because most vehicle manufacturers are examining the use of wireless local area network (LAN) technology as a mechanism for vehicles becoming participants in an ad hoc network, we will conclude this chapter with a preliminary overview of wireless LAN standards, which will be expanded upon in the next chapter.

2.1 Fundamental Concepts

In this initial section we will turn our attention to obtaining an appreciation of fundamental communications concepts. Thus, this section will acquaint us with some basic relationships that govern the operation of wired and wireless communications.

2.1.1 Powers of 10

As a refresher for readers who may be a bit rusty remembering the prefixes for the powers of 10, Table 2.1 lists four common prefixes and their meanings. In the wonderful world of both wired and wireless communications you will often encounter the term *milliwatt* (mW) when discussing communications power, while the terms *kilohertz* (kHz), *megahertz* (MHz), and *gigahertz* (GHz) are commonly associated with the operating frequencies of different communications systems.

2.1.2 Frequency

Frequency represents the term used to denote the number of periodic oscillations or waves that occur per unit time. Figure 2.1 illustrates two oscillating sine waves that are at different frequencies. The top sine wave is shown operating at one cycle per second (cps), while the bottom sine wave is shown operating at 2 cps. Note that the term *cycles per second* in general has been replaced by the synonymous term *hertz* (Hz) in honor of the German physicist.

By itself a continuously oscillating signal only tells us we have continuity when we hear it; however, it does not carry any other information. To impress information onto the signal, one must modulate or vary it. The most common types of modulation result in the amplitude, frequency, or phase of the signal being varied. For example, we could use a sine wave at 1 Hz to represent a binary 1, while a sine wave at 2 Hz could be used to denote a binary 0. This was in fact one of the earliest methods of modulation developed and is referred to as frequency shift keying (FSK) modulation.

Table 2.1 Common Prefixes of
Powers of 10

Prefix	Meaning	
Milli	1/1000	Thousandth
Kilo	1000	Thousand
Mega	1,000,000	Million
Giga	1,000,000,000	Billion

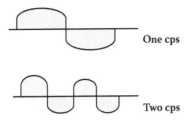

One cps

Two cps

Figure 2.1   Oscillating sine waves at different frequencies.

2.1.2.1   Signal Period

Returning to our examination of the term *frequency*, the time required for a signal to be transmitted over a distance of one wavelength is referred to as the period of the signal. The period represents the duration of the cycle and can be expressed as a function of the frequency. That is, if T represents the period of a signal and f is the frequency, then the relationship between T and f becomes

$$T = 1/f \text{ or } f = 1/T$$

Based upon the preceding, note that the sine wave shown at the top portion of Figure 2.1, whose signal period is 1 s, has a frequency of 1/1, or 1 Hz. Similarly, the second sine wave whose period is 0.5 s has a frequency of 1/0.5, or 2 Hz. Thus, as the period of a signal decreases, its frequency increases.

2.1.2.2   Wavelength

The period of an oscillating signal is also referred to as its wavelength, with the Greek character lambda used. The wavelength in meters of a signal can be obtained by dividing the speed of light (3×10^8 m/s) by the signal's frequency in Hz. That is,

$$\text{Lambda (m)} = 3 \times 10^8/f \text{ (Hz)}$$

2.1.3   Bandwidth

Bandwidth represents a measure of the width of a range of frequencies and not the frequencies themselves. For example, if the lowest frequency that can be used in a frequency band is f_1 and the highest is f_2, then the available bandwidth is $f_2 - f_1$. In telephone operations the bandwidth of the average human is considered to be 20,000 Hz, ranging from 0 to 20,000 Hz. However, the telephone company uses filters to remove frequencies below 300 Hz and above 3300 Hz, resulting in 3000 Hz of bandwidth used for voice communications.

2.1.4   Power Measurements

The development of the telephone resulted in a need to define the relationship between the received power of a signal and its original or input power. Initially, this relationship was the bel (B), named in honor of Alexander Graham Bell, the inventor of the telephone.

2.1.4.1   The Bel

Named after the inventor of the telephone, the bel (B) uses logarithms to the base 10 to express the ratio of power transmitted to power received. The resulting gain or loss of a circuit is given by the following formula:

$$B = \log_{10} P_o/P_I$$

where B is the power ratio expressed in bels, P_o is the received or output power, and P_I is the transmitted or input power.

The rationale for the use of logarithms to the base 10 corresponds to the manner by which humans hear sounds. That is, our audio capability perceives sound or loudness on a logarithmic scale. For example, if you estimate, based upon your hearing, that a signal doubled in its loudness, the transmission power actually increased by approximately a factor of 10.

A second reason for the selection of logarithms for use in power measurements results from the fact that changes to a signal in the form of signal loss due to resistance or signal gain due to the use of an amplifier are additive. Thus, the ability to add and subtract when performing power measurements based on a log scale considerably simplifies computations. For example, a 10-B signal that encounters a 5-B loss and is then passed through a 20-B amplifier results in a signal strength of $10 - 5 + 20$, or 25 B.

2.1.4.2   Log Relationships

There are two log relationships worth noting that can be used to simplify power measurement computations. First, you can note that the logarithm to the base 10 ($\log_{10}$) of a number is equivalent to determining how many times 10 is raised to a power to equal the number. For example, $\log_{10} 100$ is equivalent to determining how many times 10 is multiplied by itself (raised to a power) to equal 100, with the answer being 2. Similarly, $\log_{10} 1000$ is 3, $\log_{10} 1000$ is 4, and so on.

In examining the preceding equation, we can note that under normal circumstances the output or received power can be expected to be less than the input or transmitted power. When this situation occurs, the numerator in the preceding equation (P_o) will be less than the denominator (P_i). To simplify computations when this situation arises, we can use a second property associated with the use of logarithms. This second property permits us to easily resolve fractional computations, as we merely have to prefix the computation with a negative sign to flip its fraction to a whole number. That is,

$$\log_{10} 1/x = -\log_{10} X$$

Once we prefix the computation with a negative sign and flip the numerator and denominator, it becomes relatively simple to compute the log. For example, let us assume that the power received is 1/100 of the transmitted power. Then, our initial computation of the gain or loss becomes

$$B = \log_{10} 1/(100/1) = \log_{10} 1/100$$

As previously noted, $\log_{10} 1/x = -\log_{10} X$. Thus, using this relationship we obtain

$$B = -\log_{10} 100 = -2$$

Note that the computational result is negative, which indicates that a power loss occurred, and is precisely what we would expect because the received power was a very small fraction of the transmitted power. In comparison, a positive bel value would indicate a power gain, because the output power would be greater than the input power. Although your initial reaction might be dubious concerning a power gain, you need to remember that a signal flowing through an amplifier could result in this situation occurring. In a wireless LAN environment, a client transmitting to another client through an access point (AP) has its signal regenerated by the AP. If the AP has a higher power level, this situation (AP) is equivalent to an amplifier in a wired environment.

Although the bel was used for many years to categorize the quality of a transmission circuit, it gradually lost favor due to the requirement for a more precise measurement. The use of the decibel (dB) provided industry with the precise measurement it sought and, for all practical purposes, has replaced the use of the bel. Thus, let us turn our attention to the decibel, which is better known by its abbreviation, dB.

2.1.4.3   The Decibel

The decibel (dB) represents the standard used today to denote power gains and losses. As previously noted, the decibel represents a more precise measurement than the bel. This is because the dB represents 1/10 of a bel. To indicate this, we multiply the previously noted computation of the bel by 10 to obtain the computation for the decibel. That is, the power measurement in terms of decibels is computed as follows:

$$dB = 10 \log_{10} P_o/P_I$$

where dB is the power ratio in decibels, P_o is the output power or received power, and P_I is the input power or transmitted power.

To illustrate an example of the computation of a power ratio in dB, let us return to our prior power ratio computational example. In that example the received power was measured to be 1/100 of the transmitted power. Thus, the power ratio in decibels becomes:

$$dB = 10 \log_{10} 1/(100/1) = 10 \log_{10} (1/100)$$

Because $\log_{10} 1/X = -\log_{10} X$, we obtain

$$dB = -10 \log_{10} 100 = -20$$

In comparing the results of our computations for the bel and decibel for the same input and output power measurements, note that the decibel is precisely 10 times the value computed for the bel. Thus, the dB provides the ability for more precise power measurements and today is the preferred power measurement in use.

2.1.4.4   Decibel above 1 mW

The terms bel (B) and decibel (dB) represent a ratio or comparison between two values, such as input and output power. Although they are important tools, they are not useful for comparing two circuits because they do note specify a common input power level. Thus, for comparing two or more circuits we would want to inject the same amount of power into each and observe the level of received power. In the wonderful world of telecommunications testing, a 1-mW signal occurring at a frequency of 800 Hz is used in North America. To ensure that you do not forget that testing occurred with respect to a fixed 1-mW signal, the term *decibel-milliwatt* (dBm) is used.

Output power with respect to a 1-mW test tone is computed in dBm as follows:

$$dBm = \log_{10} \text{output power}/1 \text{ mW input}$$

Table 2.2   Relationship of Watts and dBm

Power in Watts	Power in Decibel-Milliwatts
0.001 mW	–30 dBm
0.01 mW	–20 dBm
1 mW	0 dBm
1 W	30 dBm
1 kW	60 dBm
1 MW	90 dBm

Note we use the term *dBm* to remind us that the output power measurement occurred with respect to a 1-mW test tone. Although the term *decibel-milliwatt* is used in most literature, in actuality *dBm* means decibel above 1 mW because the output or received signal is based upon the input of a 1-mW signal. Thus, 10 dBm more correctly represents a signal 10 dB above or bigger than 1 mW, while 20 dBm represents a signal 20 dB above 1 mW, and so on.

One interesting power relationship concerns a 30-dBm signal. A 30-dBm signal is 30 dB or 1000 times larger than a 1-mW signal. Thus, 30 dBm is equal to 1 W. We can use this relationship of 30 dBm being equal to 1 mW to construct a watts–to–decibel-milliwatt conversion table (Table 2.2).

Most of the entries in Table 2.2 should be self-explanatory; however, let us review them to ensure we are all on the same path. Let us start our review with the third line in the table, where 1 mW is shown as equivalent to 0 dBm. Because

$$dBm = 10 \log_{10} P_o/P_I$$

the only way to achieve 0 dBm is for P_o to equal P_I. Thus, 0 dBm must have an output power of 1 mW. Once we understand that 0 dBm is equivalent to 1 mW, then the other entries in the table are easy to understand. For example, 1 W is 1000 times greater than 1 mW. Because 30 dBm represents a signal 1000 times that of a 1-mW signal, 1 W is equal to 30 dBm. Similarly, 1 kW is 1000 times greater than 1 W and 1 MW is 1000 greater than 1 kW. Thus, we need to add 30 dBm for each, resulting in 1 kW being equal to 60 dBm and 1 MW being equal to 90 dBm. The only remaining entries to review are the first two. Because 0.001 mW is 1/1000 of the input power of 1 mW, we obtain

$$dBm = 10 \log_{10} 0.001/1$$

Thus, –30 dBm is equal to 0.001 mW of power. Similarly, 0.01 mW requires an output 1/100 of 1 mW, which is equal to –20 dBm.

2.1.4.5   The Decibel Isotropic

Because wireless devices communicate via the use of antennas, another metric that warrants our attention is decibel isotropic, abbreviated dBi. This metric is used to define the gain of an antenna relative to a hypothetical antenna that radiates output uniformly in all directions. This uniform radiating antenna only exists in theory and is known as an isotropic antenna. Thus, dBi represents a measurement of how much better an antenna is in comparison to an antenna that transmits signals equally in all directions.

The computation of dBi is based upon the decibel, resulting in the gain (G) of an antenna. Specifically,

$$G = 10 \log_{10} (I_A/I_i)$$

where G is the gain of the antenna in dBi, I_A is the electromagnetic field of intensity measured in microwatts per square meter (mW/m²) generated by antenna A, while I_i represents the electromagnetic field of intensity produced by an isotropic antenna and similarly measured in microwatts per square meter, with both measurements occurring at the same distance from the antennas.

Because the dB is commonly used in communications, it is often helpful to have access to a table of dB and power ratio values. Table 2.3 will assist readers in any computations they may have to perform; decibels range from 0 to 100 dB with the equivalent power ratio for each dB entry. Because dB values are algebraic, you can add or subtract one value from another to obtain a desired value. For example, from Table 2.3 the ratio for 10 dB is 10, while the ratio for 20 dB is 100, because doubling the dB increases the power ratio by a factor of 10. Working up from 10 dB, +3 dB is equivalent to a power ratio of 2 × 10, or 20. Thus, 13 dB is equal to a power ratio of 10 + 20, or 30. Similarly, adding another 3 dB (10 + 3 + 3) for a total of 16 dB results in a power ratio of 50 (10 + [2 × 10] + [2 × 10]). If we added another 2 dB for a total of 18, our power ratio would become 65.8 (10 + [2 × 10] + [2 × 10] + [1.58 × 10]).

2.1.4.6   Considering Power Limits

In a wireless LAN environment Federal Communications Commission (FCC) regulations restrict transmission to 36 dBm (4 W) of equivalent isotropic radiated power (EIRP) in the 2.4-GHz frequency band. Because transmitter power and antenna gains are cumulative, you need to consider both to stay within legal limits. Table 2.4 provides a summary of the relationship among power injected into an antenna, antenna gain in dBi, and EIRP in dB.

Table 2.3 Decibel Reference Table

dB	Power Ratio
0	1.0
0.5	1.12
1.0	1.26
1.5	1.41
2.0	1.58
3.0	2.00
4.0	2.51
5.0	3.16
6.0	3.98
7.0	5.01
8.0	6.31
9.0	7.94
10	10.00
15	31.6
20	100
25	316
30	1000
40	10,000
50	100,000
60	1,000,000
70	10,000,000
80	10,000,000
90	10,000,000
100	100,000,000

2.1.4.7   Antenna Selection

When the distance between wireless LAN stations increases, you can either add repeaters or use directional antennas to better support communications. For example, a single-element antenna may provide a gain of 6 dBi, whereas a specialized parabolic antenna could provide a gain well over 24 dBi. However, because the EIRP maximum is fixed at 36 dBm, you may need to lower the transmit power when using certain types of high-gain antennas. For example, if transmit power is 27 dBm (500 mW), you can only use an antenna with a gain of 9 dBi or less, as any gain over 9 dBi would result in an EIRP greater than 36. Of course, you could

Table 2.4 Legal Relationship among Power Injected into an Antenna, Antenna Gain, and EIRP in the 2.4-GHz Band

Power at Antenna (dBm/W)	Antenna Gain (dBi)	EIRP (dBm)
30 dBm (1 W)	6	36
27 dBm (500 mW)	9	36
24 dBm (250 mW)	12	36
21 dBm (125 mW)	15	36
18 dBm (62.5 mW)	18	36
15 dBm (31.25 mW)	21	36
12 dBm (15.125 mW)	24	36

lower the level of transmit power, but doing so would defeat the purpose of acquiring a high-gain antenna.

If you have low-power devices, you can consider the use of an antenna with multiple elements, referred to as an array antenna. One popular type of antenna is a four-element array antenna, which can be used to enhance transmission distance by generating radio frequency (RF) energy in a particular direction. Such antennas are referred to as tuned element array antennas. By using different frequencies for each element and changing their transmission phase, the antenna becomes a phased-array antenna, with operating characteristics similar to those of Distant Early Warning (DEW) phased-array radar. That is, such antennas become highly directional and their use in wireless LANs may occur by the time you read this book.

2.1.4.8   Receiver Sensitivity

The sensitivity of an antenna has a considerable bearing on its ability to receive a signal. Although the FCC does not place limits on receiver sensitivity, Institute of Electrical and Electronics Engineers (IEEE) standards denote receiver performance. For operations in the 2.4-GHz band, IEEE wireless equipment should have antenna sensitivity less than or equal to −80 dBm. Because 60 dB is equivalent to a ratio of 1 million to 1, −80 dBm means that the antenna should be capable of picking up 1/100,000,000 of a signal.

Wireless LANs are essentially line-of-sight transmission systems. Thus, as you might expect, their transmission distance is greater outdoors than indoors. Table 2.5 lists the transmission distances for a Linksys (now a part of Cisco) 802.11b access point. By appropriately placing repeaters, it is possible to extend transmission distances, which enables wireless devices to form a mesh network that spans a considerable distance. As we continue our journey of exploration in future chapters in this

Table 2.5   Linksys Access Point Transmission Distances

Data Rate	Outdoors	Indoors
11	50 m (164 ft)	250 m (820 ft)
5.5	80 m (262 ft)	350 m (1148 ft)
2	120 m (393 ft)	400 m (1312 ft)
1	150 m (492 ft)	500 m (1640 ft)

book, we will note how vehicles can be connected to one another in a mesh network environment that enables one vehicle to communicate with another via the relaying of information through other vehicles.

2.1.5   Signal-to-Noise Ratio

One of the more important metrics in the field of data communications is the signal-to-noise ratio, as this ratio defines the ability of a receiver to recognize a signal. In all communications systems a degree of noise exists due to the movement of electrons, power-line induction, sunspots, and the cross-modulation known as crosstalk from adjacent wire pairs in a wired network or adjacent frequencies used in a wireless network. This noise can be grouped into two basic categories: thermal and impulse. Thermal noise, such as the movement of electrons or basic radiation from the sun, can be characterized by a near-uniform distribution of energy over the frequency spectrum.

2.1.5.1   Thermal Noise

Figure 2.2 illustrates an example of thermal noise occurring over a range of frequencies. Note that thermal noise is categorized by a near-uniform distribution of energy over the frequency spectrum. Thus, for a receiver to have the ability to distinguish a signal, the signal must have a power level above the thermal noise. This

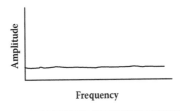

Figure 2.2   Thermal white noise is characterized by a near-uniform distribution of energy over the frequency spectrum.

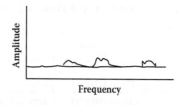

Figure 2.3 Impulse noise occurs at random times and frequencies.

means that we can consider thermal noise as representing the lower level of sensitivity of a receiver. Some other terms used as synonyms for thermal noise include *white noise* and *Gaussian noise*.

A second type of noise that affects communications results from periodic disturbances, such as electromagnetic fields generated by machinery, sunspots or solar flares, and lightning. This type of noise is referred to as impulse noise and is illustrated in Figure 2.3. Note that impulse noise consists of irregular spikes or pulses of relatively high amplitude and short duration.

2.1.5.2 Categorizing Communications

The signal-to-noise (S/N) ratio is commonly used to categorize the quality of communications transmission for both wired and wireless systems. Measured in decibels, the S/N ratio is defined as the ratio of the signal power (S) divided by the noise power (N) on a transmission medium. Although you always desire a S/N ratio above unity, because the receiver must be able to discriminate the signal from the noise, there are limits that govern the maximum signal strength. Those limits are imposed by system operators that do not want excessive power input into their systems that could adversely affect their network, as well as by the Federal Communications Commission (FCC) or another regulatory body, with limits defined by the latter designed to minimize interference between different wireless systems. After all, it would not be safe to have wireless LANs using frequencies near those used by airport control towers.

2.1.6 Transmission Rate Constraints

There are two key constraints that govern the ability to transmit information at different data rates. Those constraints are referred to as the Nyquist relationship and Shannon's law.

2.1.6.1   The Nyquist Relationship

The Nyquist relationship governs the maximum signaling capacity on a communications channel. In 1928, Henry Nyquist developed the relationship between the bandwidth (W in hertz) and the signaling rate (B in baud) on a channel as follows:

$$B< = 2\,W$$

To obtain an appreciation for the Nyquist relationship, we need to review the differences between the terms *bit rate* and *baud*. The bit rate, typically presented in bits per second (bps), represents a measurement of data throughput. In comparison, baud represents the rate of signal change, commonly expressed in hertz. When information in the form of a binary sequence of 0s and 1s is to be transmitted, an oscillating wave is varied to impress or modulate information. An example of an oscillating wave, which is referred to as a carrier, was previously shown in Figure 2.1. As mentioned earlier in this chapter, common modulation techniques include altering the amplitude, frequency, and phase of a carrier.

Because a communications channel has a fixed bandwidth defined by either the medium or a regulator, the Nyquist relationship tells us that the maximum signaling rate is limited to twice the bandwidth prior to one signal interfering with the next signal, a condition referred to as inter-symbol interference. Thus, if the signaling rate is fixed at a maximum of 2 W, then the only method to obtain a higher bit rate is to pack more bits into each signal. To illustrate this concept, let us discuss two modulation methods: frequency shift keying (FSK) and phase modulation (PM).

Under FSK each binary digit is modulated using one of two frequencies, referred to as f_1 and f_2. If we assume that all binary 1s are modulated using f_1 while all binary 0s are modulated by transmitting at f_2, this simple modulation scheme results in one bit being equivalent to one signal change, or the bit rate then equals the baud rate. Now let us assume a version of phase modulation is used where the phase of the carrier is varied to one of four positions, 0, 90, 180, or 270°. This modulation technique would enable each combination of two bits to be encoded as one signal change, as illustrated in Table 2.6. Here the bit rate is twice the signaling rate and illustrates how the data rate achieved on a channel can overcome the Nyquist relationship by packing more bits per signal change. Of course, there is an upper limit to the number of bits that can be packed into each signal change. That limit is based on the quality of the communications channel, which represents a second constraint known as Shannon's law.

2.1.6.2   Shannon's Law

In 1948, Professor Claude E. Shannon of Harvard University presented a paper concerning the relationship between coding and noise. In his paper Professor

Table 2.6 Using Phase Modulation to Pack Two Bits into Each Phase Change

Bits	Phase (degrees)
00	0
01	90
10	180
11	270

Shannon computed the theoretical maximum bit rate that could be obtained when transmitting over a channel having a bandwidth W Hz. The relationship defined by Shannon is as follows:

$$C = W \log_2 (1 + S/N)$$

where C is the capacity of a channel in bits per second (bps), W is the bandwidth in hertz, S is the transmitted power, and N is the power of thermal noise.

In 1948, a "perfect" telephone channel was considered to be one with an S/N ratio of 30 dB, which represents a value of 1000. According to Shannon's law, the maximum transmission rate achievable on a telephone channel in 1948 was

$$C = W \log_2 (1 + S/N)$$

$$C = 3000 \log_2 (1 + 10^3)$$

$$C = 3000 \log_2 (1001) = 30,000 \text{ bps}$$

It should be noted that now, approximately 60 years after Shannon's paper was presented, the maximum data rate of modems used on analog telephone lines is slightly over 44 kbps. This is a higher data rate than available under Shannon's law and results from the fact that modern modems use approximately 10 to 15 percent more bandwidth and the noise level on a telephone line has a bit (no pun intended) less noise. Concerning the latter, in 1948 all telephone lines were analog. In comparison, today the entire long-distance network is completely digital, which results in a lower amount of noise and a higher S/N ratio, which in effect permits a higher data rate to be achieved.

2.2 Radio Frequency Spectrum Allocation

Most countries have a government agency that regulates the use of the frequency spectrum. Under the provisions of International Telecommunications Union (ITU) treaties with most countries, those countries are obligated to comply with the radio frequency spectrum allocations specified by the ITU for international use. Doing so ensures that aircraft can contact control towers, satellite ground stations can receive satellite transponder signals, and cell phone subscribers can use their phones without encountering interference from other signals as country borders are approached. Although ITU treaties allow each country to allocate the domestic use of the frequency spectrum differently from international allocations, they can only do so if the domestic allocation does not conflict with a neighboring country's frequency spectrum allocation. Thus, you can expect slight to major differences in the domestic use of the frequency spectrum between countries that may become more pronounced when you travel between continents.

2.2.1 U.S. Spectrum Allocation

In the United States, the Communications Act of 1934, as revised, resulted in the authority for managing the use of the radio frequency spectrum being subdivided between the U.S. Commerce Department's National Telecommunications and Information Administration (NTIA) and the FCC. NTIA administers the frequency spectrum for federal government use. In comparison, the FCC, which is an independent regulatory agency, administers the frequency spectrum for non-federal government use. The managed radio frequency spectrum currently ranges from 9 KHz to 300 GHz and is subdivided into more than 450 frequency bands.

2.2.2 Band Nomenclature

One of the more popular methods used to categorize the frequency spectrum occurs by using the wavelength as a power of 10 metric for denoting each particular frequency band. Table 2.7 provides an example of categorizing different frequency bands based on wavelength. If you turn your attention to the top entry in the referenced table, you will note that the ultra low frequency (ULF) band can also be categorized as signals with a wavelength from 10^8 to 10^7 m. Similarly, the extremely low frequency (ELF) band represents the frequency spectrum where the wavelength varies from 10^7 to 10^5 m, whereas the very low frequency (VLF) band represents the frequency spectrum where the wavelength varies from 10^0 to 10^{13}.

A second popular method used to categorize the frequency spectrum comes from one of the well-known national regulators of the frequency spectrum, the FCC. The FCC categorizes the frequency spectrum from 0 to 400 GHz by noting the overlapping of bands that fall into a range of frequencies. Table 2.8 provides

Table 2.7   Well-Known Frequency Bands

Frequency Band	Wavelength
Ultra low frequency (ULF)	10^8–10^7
Extremely low frequency (ELF)	10^7–10^5
Very low frequency (VLF)	10^5–10^4
Low frequency (LF)	10^4–10^3
Medium frequency (MF)	10^3–10^2
High frequency (HF)	10^2–10^1
Very high frequency (VHF)	10^1–1^{-1}
Ultra high frequency (UHF)	1–10^1
Super high frequency (SHF)	10^{-1}–10^{-2}
Extremely high frequency (EHF)	10^{-2}–10^{-3}
Electro-optical frequency (EOF)	10^{-3}–10^{-8}
High-energy frequency (HEF)	10^{-8}–10^{-13}

a summary of the FCC's frequency band nomenclature, indicating the frequency range that the FCC associates with distinct frequency bands as well as certain pairs of bands. A comparison of the entries in Table 2.7 and Table 2.8 indicates that there is not a direct correspondence between the two. Thus, many times technical literature will reference the frequency range of a band or particular application rather than the band nomenclature to ensure both parties are correctly referencing the same frequency.

Table 2.8   Frequency Band Nomenclature

Frequency Band	Frequency Range
Very low frequency/low frequency (VLF/LF)	0–130 KHz
Low frequency/medium frequency (LF/MF)	130–505 kHz
Medium frequency (MF)	505–2107 kHz
Medium frequency/high frequency (MF/HF)	2107–3230 kHz
High frequency (HF)	3230–2800 kHz
High frequency/very high frequency (HF/VHF)	33–162.0125 MHz
Very high frequency/ultra high frequency (VHF/UHF)	162.0125–322 GHz
Ultra high frequency (UHF)	322–2655 MHz
Ultra high frequency/super high frequency (UHF/SHF)	2655–3700 MHz
Super high frequency (SHF)	3200 MHz–27.5 GHz
Super high frequency/extremely high frequency (SHF/EHF)	27.5–32 GHz
Extremely high frequency (EHF)	32–400 GHz

2.2.3 Applications

To obtain an appreciation for the wide range of wireless applications as well as the diverse band of frequencies allocated to those applications, Table 2.9 provides examples of 20 common and evolving wireless applications. Note that this table indicates the application and frequency band used in the United States. There may be some differences in the frequency bands as you traverse the globe.

In the United States it is worth noting that the FCC subdivided frequency into approximately 450 blocks, with each block assigned for specific applications. When we discuss collision avoidance radar (CAR) later in this chapter and in subsequent chapters we will reference millimeter wave radar systems and sensors that primarily operate in the 36- to 94-GHz frequency range. This frequency range is also used for inter-vehicle communications.

Table 2.9   Common and Evolving Wireless Applications

Application	Frequency
AM radio	535–1635 KHz
Analog cordless phone	44–49 MHz
Television	54–88 MHz
FM radio	88–108 MHz
Television	174–218 MHz
Television	470–806 MHz
Wireless data	700 MHz
RF wireless modem	800 MHz
Cellular	806–890 MHz
Digital cordless	900 MHz
Personnel communications	929–932 MHz
Satellite telephone uplink	1610–1626.5 MHz
Personnel communications	1850–1990 MHz
802.11/11b/g wireless LAN	2.4–2.4835 GHz
Satellite telephone downlink	2483.5–2500 GHz
Large dish satellite TV	4–6 GHz
802.11a wireless LAN	5.15–5.35 GHz; 5.725–5.825 GHz
Small dish satellite TV	11.7–12.7 GHz
Wireless cable TV	28–29 GHz
Inter-vehicle communications	36–94 GHz

2.3   Radar Operations

Radar represents an acronym for radio detection and ranging, a technology that involves directing a beam of microwave energy at a location from which a target may be approaching or receding. If the target is hit by the beam, a portion of its energy is reflected and received by the radar unit that transmitted the signal. The reflected signal is shifted in frequency by an amount proportional to the speed of the target, making it possible to determine the speed at which a vehicle is approaching or leaving the source of the radar. The shift in frequency is referred to as the Doppler effect.

Although radar was first used by the military during World War II, its operation resulted in the development of the radar gun, radar-controlled camera, and other devices, to the chagrin of motor vehicle operators who received speeding tickets either from the partner of the police officer operating the radar gun, located a mile down the road, or in the mail due to a photograph of the vehicle license plate being taken for exceeding a certain threshold. In this section we will first examine police traffic radars because they are commonly employed to detect speeders. Once this is accomplished, we will turn our attention to millimeter radar, which is gradually being incorporated into vehicles to prevent collisions and support a variety of safety features.

2.3.1   Police Radar

Since the first radar unit was developed in 1947 for use by the state police, radars have evolved so that they can operate from a stationary position of a moving patrol vehicle. All moving radars measure oncoming traffic, while some devices can also measure receding traffic, same-lane traffic, and track either one or two targets. The latter occurs by setting the device to measure either the strongest reflection, to indicate the closest or largest targets, or the echo time, to indicate the fastest targets.

Police radar operates in a distinct frequency band. Table 2.10 lists the frequency, tolerance, and frequency range for an obsolete (S-band) and four existing

Table 2.10   Radar Band Characteristics

Band	Frequency Range (GHz)
S	2.400–2.4835
X	10.500–10.550
Ku	13.450
K	24.025–24.225
K	24.050–24.250
Ka	33.400–36.000

radar bands (X, Ku, K, Ka), although for one band (Ku) no equipment has been sold in the United States.

2.3.1.1   S-Band Radar

The S-band police radar dates to 1947, when it was introduced for use by Connecticut state police. Because this radar operated in the same frequency range as many microwave ovens and the 2.4000- to 2.4835-GHz frequency band was assigned to industrial, scientific, and medical (ISM) unlicensed operations, S-band radars are now obsolete.

2.3.1.2   X-Band Radar

X-band police radar dates to the mid-1960s. This radar operates at 10.525 GHz ± 25 MHz and provides better performance in all weather conditions because there is less signal attenuation in bad weather than K- or Ka-bands. In Europe, some countries use X-band traffic radars at 9.41 or 9.90 GHz.

2.3.1.3   Ku-Band Radar

Although the Federal Communications Commission (FCC) allocated 13.45 GHz in the Ku-band for traffic radar in the United States, no such radars were sold or used in the United States. However, some European countries use Ku-band 13.45-GHz traffic radar.

2.3.1.4   K-Band Radar

K-band radar dates to 1976 and operates on a single frequency of 24.125 or 21.150 GHz with a tolerance of ±100 MHz. K-band radar has a wider beam than Ka-band radar but a more narrow beam than an X-band radar.

2.3.1.5   Ka-Band Radar

The Ka-band represents the most recent allocation of frequency for traffic radar use. Initially, the FCC allocated the 34.2- to 35.2-GHz frequency spectrum for traffic radar use. In 1992 the FCC expanded the Ka-band for traffic radar use to 33.4 to 36.0 GHz.

Ka-band radar has a narrower beam than X- or K-band radars and operates with a tolerance of ±100 or ±50 MHz. What makes the Ka-band radar distinct from other traffic radar is a wideband version that operates on one frequency for a fraction of a second and then hops to another frequency, hopping through the

34.2- to 35.2-GHz frequency spectrum using either 13 channels (2600/200) or 26 channels (2600/100).

Here frequency hopping can be viewed as a countermeasure aimed at defeating vehicles operating over the speed limit that have radar detectors.

2.3.2   Types of Radar

There are several types of radar systems that can be classified by the frequency they operate on and how they operate. Concerning the latter, some microwave radars constantly transmit. In comparison, other types of radar only transmit on operator command. An example of the former might be radar on the DEW line designed to provide a warning if Russian planes flew toward North America over the polar ice cap, while an example of the latter would be a handheld radar gun operated by the state police to ensure compliance with traffic speed limits. A third version or type of radar operates only periodically, transmitting a pulse every few seconds to obtain speed measurements of an oncoming vehicle.

2.3.2.1   Laser Radar

The previous types of radar discussed were all based upon microwave technology. A second type that warrants a brief mention is laser radar, which uses a laser to generate a light pulse or series of pulses in the upper infrared (IR) band and has extremely narrow beams in comparison to microwave-based radar. Although laser radar can measure speed and range, it requires an exact aim and cannot be used within a patrol vehicle from behind glass. In addition, laser radar's detection capability is considerably decreased by fog, rain, dust, smoke, and humidity. Due to these limitations, automobile manufacturers have essentially ruled out the use of laser technology.

2.3.2.2   Collision Avoidance Radar

Collision avoidance radar (CAR) can be considered to represent a type of inter-vehicle communications as a vehicle uses radar to scan for other vehicles and objects to assist the driver in preventing a collision. Because very high frequencies provide a line-of-sight transmission capability, millimeter wave radar and radar sensors for CAR operate in the 36- to 94-GHz frequency range, carefully avoiding frequencies used by police radar to prevent interference. Radar sensors for maintaining a distance between vehicles are mounted on the back of vehicles, enabling a "middle" vehicle to measure the range to the vehicle in front. The middle vehicle is equipped with a sensor to note the position of the vehicle behind, which may not be equipped with a sensor. Because the sensor is passive, it is more economical than placing millimeter wave radar on the front and back of vehicles. Unfortunately, it would be a significant effort to equip all existing vehicles with a sensor, which makes this author believe that to be successful, collision avoidance radar may be required to operate on both the front and rear of vehicles. Thus, although practical for use in military convoys, CAR may not be suitable for commercial use in public vehicles.

2.4   IEEE Wireless LANs

In this section we will focus our attention upon a series of IEEE wireless LAN standards. The rationale for understanding IEEE wireless standards results from the fact that many vehicle manufacturers and standards bodies are examining the use of wireless LAN technology to provide the basis for ad hoc networking that will allow vehicles to automatically join and exit communications networks. When joined to an ad hoc network, inter-vehicle communications will allow vehicles to exchange safety, traffic, and other types of information.

2.4.1   IEEE Standards

In the United States the American National Standards Institute (ANSI) delegated responsibility for the development of LAN standards to the IEEE. As a result of this delegation, the IEEE initially developed standards for wired Ethernet and token ring LANs during the 1980s. Approximately 20 years later the IEEE developed its first wireless LAN standard, which is referred to as the 802.11 standard.

2.4.1.1   The 802.11 Standard

The first wireless LAN standard developed by the IEEE was 802.11. This standard defined the use of three physical layers for wireless communications: infrared, frequency-hopping spread spectrum (FHSS), and direct-sequence spread spectrum (DSSS). Although vendors developed products that used FHSS and DSSS for wireless LANs during the late 1990s, to the best of this author's knowledge no products were ever developed to follow the IEEE 802.11 infrared communications standard.

2.4.1.1.1   FHSS

Under frequency-hopping spread spectrum (FHSS) a station transmits for a small period of time at one frequency, with the period referred to as dwell time, and then hops to a different frequency to continue communications. The frequency-hopping algorithm is known to each LAN station, enabling each station to adjust its transmitter or receiver according to its mode of operation. One of the more interesting aspects of FHSS is the fact that its origin dates to the actress Hedy Lamarr, who suggested the technique to the U.S. War Department during WWII as a transmission security mechanism.

2.4.1.1.2   DSSS

Direct-sequence spread spectrum (DSSS) represents a second transmission technique developed by the military to overcome enemy jamming. Under DSSS a spreading code is applied to each bit to spread the transmission. At the receiver a "majority rule" rule is applied. That is, if the spreading code is five bits and the bits received were 10110, because three bits are set the receiver would assume the correct bit is a 1. Under the IEEE 802.11 standard an 11-bit spreading code is employed.

2.4.1.1.3   Frequency Use

The original 802.11 standard operates in the unlicensed ISM band from 2.4 to 2.4835 GHz. With the exception of the 802.11a standard that operates in the 5.15- to 5.35-GHz and 5.725- to 5.825-GHz frequency bands, all other IEEE LANs operate in the lower-frequency band. Because high frequencies attenuate more rapidly than lower frequencies, the 802.11 standard has a shorter transmission range than other members of the wireless LAN family of standards.

2.4.1.1.4   LAN Utilization

The initial use of 802.11 wireless LANs was limited due to their relatively low data rate. This was because each of the three physical layers was only defined for operations at 1 and 2 Mbps. Recognizing the need for a higher data transmission rate resulted in the IEEE initially developing two extensions to the basic 802.11 standard. Those extensions are known as the 802.11a and 802.11b standards.

2.4.1.2   *The 802.11a Standard*

The IEEE 802.11a standard defined a series of new modulation methods that enable data transmission rates up to 54 Mbps. The higher data rates are obtained by the use of orthogonal frequency division multiplexing (OFDM), a technique in which the frequency band is divided into subchannels that are individually modulated. The IEEE 802.11a standard defines operations in the 5-GHz frequency band. This means equipment supporting the standard is not backward compatible with the basic 802.11 standard because that standard defines operations in the 2.4-GHz frequency band. In addition, because high frequencies attenuate more rapidly than low frequencies, 802.11a wireless LAN stations have a shorter range than stations operating in the 2.4-GHz band. This in turn requires an organization to deploy more access points to obtain a similar geographical area of coverage than would be required via the use of access points operating in the 2.4-GHz band.

2.4.1.3   The 802.11b Standard

The second extension to the basic IEEE 802.11 standard is 802.11b. Under the IEEE802.11b standard DSSS was used with two new modulation methods to provide a data transfer rate of 11 and 5.5 Mbps. The 802.11b standard also provides compatibility with 802.11 DSSS equipment operating at 2 or 1 Mbps. To provide this compatibility, the IEEE802.11b standard specifies the use of the 2.4-GHz frequency band.

2.4.1.4   The 802.11g Standard

A comparison of the IEEE 802.11a and 802.11b extensions to the 802.11 standard indicates advantages and disadvantages associated with each. Although the 802.11a standard provides a higher data transfer rate, its use of the 5-GHz frequency band results in a shorter transmission distance. Similarly, in a reverse manner, the IEEE 802.11b standard provides a greater transmission distance but lower data rate than obtainable from the use of 802.11a-compatible equipment. By combining the modulation method used in the 802.11a standard with the frequency band employed by the 802.11b standard, the IEEE provided a mechanism to extend both the data rate and transmission range of wireless LANs, resulting in the 802.11g standard. To provide backward compatibility with the large base of 802.11b equipment, the 802.11g standard also supports DSSS operations at 11, 5.5, 2, and 1 Mbps. Thus, the relatively new IEEE802.11g standard can be considered to represent a dual standard because it provides 802.11b compatibility.

2.4.1.5   The 802.11n Standard

The 802.11n standard actually represents an emerging standard that will probably be finalized in late 2007 or early 2008. This evolving standard provides a significant increase in transmission rates due to its support of multiple-input, multiple-output (MIMO) technology. MIMO exploits a radio wave phenomenon called *multipath* in which transmitted information bounces off objects, reaching a receiving antenna multiple times via different routes with slightly different delays.

Normally, multipath results in the distortion of a transmitted signal. However, through a technique referred to as *space-division multiplexing* the transmitting device splits a data stream into multiple parts known as *spatial streams*, with each spatial stream transmitted through separate antennas to correspond with multiple antennas at the receiver. The current 802.11n draft standard specifies that up to four spatial streams can be supported that can theoretically result in a maximum data rate up to 600 Mbps. This high data rate depends on not only four spatial streams, but also the use of a 40-MHz channel instead of the 20-MHz channels specified for 802.11a, 802.11b, and 802.11g equipment.

Under the draft 802.11n, standard equipment can optionally use 20- or 40-MHz channels and operate in either the 2.4- or 5-GHz frequency band. Through the use of multiple antennas, which improves efficiency by allowing transmission bursts of multiple data packets between overhead communications and a reduction in inter-frame spacing that results in a shorter time delay between transmissions, users can expect to achieve at least a 100-Mbps data transfer ability. Although the higher data transfer can be important for home and business wireless LAN applications, its use in vehicle operations more than likely represents an excessive data transfer capability.

Chapter 3

Communications Technologies

In this chapter we will expand our knowledge of communications based upon the foundation presented in the previous chapter. We will describe and discuss the strengths and weaknesses of cellular, Bluetooth, WiFi, and satellite transmission systems with respect to their actual and potential uses in vehicles. Once this is accomplished, we will examine ad hoc networking and discuss how data can be routed between vehicles via an ad hoc network. To accomplish this, we will discuss the emerging IEEE 802.11s standard as a precursor to discussing mobile ad hoc networking, which is referred to as MANET.

3.1   Transmission Technologies

There are four basic methods that can be used to provide a communications capability within and between vehicles. Those communications technologies include the use of a communications protocol that operates on the wiring installed within a vehicle as well as three wireless technologies. Because we will cover in detail in future chapters in this book several communications protocols used in intra-vehicle communications, we will focus our attention upon wireless transmission systems in this section. Beginning with cellular communications, we will examine the use and advantages and disadvantages of Bluetooth, WiFi, and satellite-based communications.

3.1.1 Cellular Communications

There are two major types of cellular communications used in the United States for voice communications: Global System for Mobile (GSM) and code division multiple access (CDMA). In the following two sections we will become familiar with each.

3.1.1.1   GSM

The Global System for Mobile (GSM) communications represents the most popular communications technology used in cellular phones on a worldwide basis. Both its signaling and speech channels are digital, which in effect represents a second generation of technology when one considers the fact that the Analog Mobile Phone Service (AMPS), which represents the initial cellular service, is an analog technology that has been shut down at most, if not all, of its original service locations.

GSM represents a cellular network, requiring mobile phones to scan for a cell signal to connect to the GSM network. As a subscriber moves from one cell to another, a handoff process occurs, allowing the subscriber to continue his or her network connection in the new cell.

3.1.1.1.1   Frequency Use

In the United States and Canada GSM networks operate in the 850- and 1900-MHz frequency bands. In Europe, parts of Asia, and the rest of the world, GSM networks typically operate in the 900- and 1800-MHz frequency bands. The major reason for the use of different frequency bands results from the fact that GSM evolved in Europe, and when it began operations in North America, the 900- and 1800-MHz frequency bands were already allocated for a different type of communications. Thus, when you obtain a new mobile phone that you want to use on a worldwide basis, you should opt for a tri-band GSM phone that supports three frequency bands to obtain a worldwide calling capability.

The 900-MHz frequency band uses 890 to 915 MHz for uplink (cell phone to cell tower) communications, while downlink (cell tower to cell phone) uses frequencies from 935 to 960 MHz. The 25-MHz bandwidth is subdivided into 124 carrier frequency channels, each spaced 200 kHz apart. Through the use of time division multiplexing (TDM), eight speech channels are placed on each radio frequency channel, allowing 124*8, or 992, simultaneous conversations to be supported within a GSM cell; however, as we will shortly note, there are four different types of GSM cells.

3.1.1.1.2   Power Limitations

The transmission power of GSM cell phones is limited to a maximum of 2 W in the 850/900 frequencies and up to 1 W in the 1800/1900 frequencies. Thus, GSM is similar to other cellular technologies in that it is a relatively low-power device. This means that communications depends upon cell towers being located either in close proximity in urban areas, where buildings obstruct transmissions, or within at most approximately 20 to 25 miles in rural areas, where the curvature of the Earth becomes an important consideration for communications.

3.1.1.1.3   Network Structure

Figure 3.1 illustrates the basic structure of a GSM network. The cell or mobile phone communicates with a base station subsystem (BSS), which consists of one or more base stations and their controllers, and a network and switching subsystem, which connects to the public switched telephone network (PSTN) to route calls over that network as well as coordinate the handoff and entry of mobile phones as they approach and enter or exit cells. While the BSS and NSS are included in all GSM networks, a General Packet Radio Service (GPRS) core network that allows packet-based Internet connections is optional; however, it is now incorporated into most GSM networks.

3.1.1.1.4   GSM Cells

There are four different cell sizes that can be supported in a GSM network — macro, micro, pico, and umbrella — with the area of coverage of each cell varying based upon the manner by which it is implemented. A macro cell represents one where the base station antenna is installed on the roof of a building or on a mast that towers above the average rooftop level. In comparison, a micro cell represents one where the base station antenna's height is below the average rooftop level. Thus, a macro cell is usually used in rural areas and provides a longer transmission distance than a micro cell, which is commonly used in built-up urban areas.

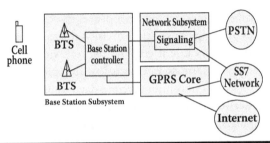

Figure 3.1   GSM network structure.

Pico cells are very small cells whose diameter may be a few dozen to a hundred meters. If you visited a casino in Las Vegas and while waiting in the registration line used your cell phone and experienced an awesome connection quality, the casino probably had several pico cells in its building, one of which serviced your call. Because the transmission to the cell tower was all indoors and of relatively short distance, you experienced a very high quality connection. The fourth type of cell is an umbrella, which is used to cover a region of smaller cells and, similar to some insurance advertisements, fills in gaps in coverage; however, instead of insurance, the umbrella cell fills in the gaps of coverage between cells.

The actual cell radius of coverage depends upon several factors, to include the antenna height, antenna gain, and propagation conditions. Typically the longest radius of coverage is 35 km, or 22 miles, in rural areas, while indoor-based pico cells may have a radius of coverage in the tens of meters.

3.1.1.1.5 Voice Compression

Originally GSM used a member of the family of linear predictive coding (LPC) voice digitization methods to reduce the data rate of a voice conversation to 13 kbps. In comparison, the telephone company uses pulse code modulation (PCM) to digitize voice at 64 kbps. Today, updates to GSM voice digitization provide PCM-like quality at data rates as low as 12.2 kbps.

3.1.1.1.6 Data Transmission

GSM provides subscribers with two popular methods to transmit data, referred to as GPRS and EGDE. In this section we will obtain an overview of each method.

3.1.1.1.6.1   GPRS — General Packet Radio Service (GPRS) represents a low-speed mobile data service available to GSM phone subscribers. Because the technology falls between the second and third generation of mobile phone technology, GPRS is often referred to as a 2.5G service.

GPRS, as its name implies, represents a packet-switched communications service. GPRS operates by using unused time division multiple access (TDMA) channels in a GSM network. Because it employs packet switching, it allows multiple users to share the same transmission channel, with data flowing on the channel only when a phone needs to transmit or receive data. Thus, the full bandwidth of a channel can be dedicated to users that are sending data while other subscribers that are reading the contents of their screen, scrolling through data, or munching on a candy bar are not using the service. Thus, Web browsing, transferring e-mails, and even instant messaging benefit by sharing the GPRS bandwidth.

Data Transfer Rates — The operational rate of GPRS depends upon the activity of the cell the subscriber is located in. This results from the fact that as dedicated voice and data channels are set up by mobile phones, the available bandwidth for use by GPRS for data is reduced. Although the theoretical maximum data transfer for GPRS is 171.2 kbps, an average data transfer between 30 and 80 kbps is probably more realistic in urban areas.

Device Classes — There are three classes of GPRS devices: A, B, and C. A Class A device can be connected to GPRS and GSM (voice) and use both services at the same time. In comparison, a Class B device can be connected to both; however, it can only use one service at a time. Last but not least, a Class C device can be connected to either GPRS or SMS; however, it must be switched manually between services.

Operation — Due to the relatively low data rate and cost of GPRS, its use is in decline. The data rate rarely exceeds that of an analog modem used on the PSTN, while its cost, which is usually expressed in dollars per Mbyte, can rapidly increase if the mobile phone subscriber surfs the Web. Thus, for GSM users the current preferred method of data transmission is based upon the use of EDGE.

3.1.1.1.6.2   EDGE — EDGE, which represents an acronym for Enhanced Data Rates for GSM Evolution, can be viewed as a digital mobile phone technology enhancement to GSM networks. EDGE is a superset to GPRS and can be easily added to GSM networks operating GPRS.

Operation — EDGE supports nine modulation and coding methods, of which five use a version of phase shift keying (PSK) in which three bits are encoded into a single phase change. Coding three bits into a single phase change triples the data rates obtainable, making it possible to achieve data rates up to 236.8 when four time slots are used and a maximum data rate of 473.6 kbps if eight time slots can be used.

Currently EDGE is supported by GSM operators in North America. Other GSM operators view a third-generation data service referred to as the Universal Mobile Telecommunications System (UMTS) as a more robust data transfer network. Under UMTS, a 5-MHz channel is used to provide a higher data transfer capability that can range from up to 384 kbps in highly mobile applications to up to 2 Mkbps in a stationary environment. Although UTMS capability is available in many European and Asian areas, its high cost and use of wideband code division multiple access (WCDMA) as its underlying technology resulted in EDGE being the preferred data transfer method for GSM network operators.

3.1.1.2   CDMA

Code division multiple access (CDMA) represents a second popular technology for cell phone networks in the United States, as well as other locations around the globe. CDMA can be viewed as a combination of multiplexing and data coding that enables more telephone calls to flow over a cell than earlier analog AMPS and digital TDMA-based systems.

3.1.1.2.1   Evolution

CDMA was pioneered by Qualcomm, Inc., and was known as IS-95, referencing an interim standard of the Telecommunications Industry Association (ISA). IS-95 can be considered to represent a second-generation cellular system that was superseded by the IS-2000 standard. Both IS-95 and IS-2000 use a 1.25-MHz shared channel, and the latter is now used by Sprint, Nextel, and Verizon Wireless. Similar to GSM networks, operators that support CDMA can provide a packet data transmission capability referred to as EVDO.

3.1.1.2.2   EVDO

Evolution-Data Optimized (EVDO) dates to 1999, when Qualcomm, the developer of CDMA, designed a high-data-rate technology for use over CDMA networks. By 2004, Alltel, the Sprint-Nextel Corporation, and Verizon Wireless were operating EVDO networks in the United States. By the end of 2006, Sprint and Verizon were advertising EVDO coverage in over 100 cities in the United States that allows subscribers to insert an EVDO PC card into their laptop computer and access the Internet at data rates up to 784 kbps, even though the theoretical maximum data rate is up to approximately 3 Mbps.

3.1.1.2.3   Cell Phone Data Transmission

In addition to PC cards, some cell phones, to include those manufactured by LG, Motorola, Nokia, and Palm, are EVDO enabled. An EVDO-equipped cell phone can be plugged into a laptop or another device to access the Web directly.

3.1.1.2.4   Data Applications

The integration of a cell phone into the command center or navigation system of a vehicle and appropriate software will enable the larger screen in the dashboard to be used to display data. This in turn would enable cell phone subscribers to surf the Internet without being restricted to the handful of sites that support the Wireless Application Protocol (WAP), which turns a cell phone into a mini-browser. Although a cell phone subscriber can use a WAP cell phone to access many Internet financial and other types of sites, the resulting display can be considered awkward when compared to accessing a Web site with a fully capable browser.

Other potential intra-vehicle applications that can be supported by an integrated cell phone include receiving traffic reports, requesting alternate routing when faced with traffic buildup, and locating services such as hotels, motels, vehicle repair shops, and gas stations. Although all of these services are available through the use of many navigation systems, several of these systems depend upon a built-in DVD loaded disk, which can be dated, while information received via a cell phone could be more up to date. In fact, cell phones could be integrated into a vehicle's navigation system to search for updates when the driver or passenger initiates a navigation request, such as for a list of nearby restaurants.

3.1.1.2.5   Advantages and Disadvantages

There are several advantages and disadvantages associated with the integration of cell phones into a vehicle's electronics system. Concerning advantages, the cell phone is almost ubiquitous. In addition, its integration can be viewed as a safety feature, because its integration into a vehicle's speaker system and hands-free operation reduce the need for a driver to fiddle with the phone and possibly lose control of the vehicle.

One disadvantage of cell phones is that coverage is mainly limited to urban areas and along major roadways. Thus, persons traveling in rural locations may lose the ability to obtain cell phone service. Another disadvantage of integrating the cell phone into a vehicle's electronics, such as its DVD or command center, is the fact that many functions that could be performed via the use of a cell phone are already available via other systems. For example, traffic delays can be noted via tuning a vehicle's radio to a predefined frequency listed on some traffic signs or obtained by a subscription to a service that downloads information directly into a vehicle's navigation system. Now that we have an appreciation for several cellular voice and data transmission methods and the advantages and disadvantages associated with their integration into a vehicle's electronics, let us turn our attention to another communications technology: Bluetooth.

3.1.1.3   Bluetooth

Bluetooth represents a wireless technology for relatively short transmission distances that are technically referred to as a Personal Area Network (PAN). This technology was named after a 10th-century king of Denmark, Harold Bluetooth, who according to legend engaged in diplomacy that resulted in warring parties negotiating with each other. Thus, the developers of the wireless PAN technology selected the name Bluetooth in recognition of its goal to allow different devices to communicate with one another.

Table 3.1   Bluetooth Device Classes

Class	Maximum Permitted Power (mW)	Maximum Permitted Power (dBm)	Range (m)
Class 1	100	20	100
Class 2	2.5	4	10
Class 3	1.0	0	1

3.1.1.3.1   Power and Range

As low-power devices, Bluetooth-capable products have a range between 1 and 100 m, with the range dependent upon the power. Bluetooth devices operate in the 2.4-GHz frequency band, and their transmission power levels fall into one of three classes, as indicated in Table 3.1. If you use a modern cell phone or digital camera that has Bluetooth compatibility, you can use a Bluetooth mobile phone handset for hands-free cell phone operation and upload pictures from you camera to a PC that has either a built-in Bluetooth capability or a Bluetooth USB adapter plugged into one of the computer's USB ports. Other devices that may have a built-in Bluetooth capability include PDAs, laptops and notebooks, printers, and video game consoles.

3.1.1.3.2   Versions

The first version of Bluetooth was 1.0, and a minor revision, 1.0B, had several problems, to include difficulty in exchanging data. Bluetooth 1.1 corrected most of the errors in the 1.0B specification. All three versions as well as Bluetooth 1.2, which added an adaptive frequency-hopping scheme to improve resistance to frequency interference by avoiding frequencies in use, operate at a maximum speed of approximately 1 Mbps. Bluetooth 2.0, which is backward compatible to versions 1.X, includes an enhanced data rate that extends the maximum transmission rate to approximately 3 Mbps. In addition, Bluetooth 2.0 has a lower power consumption as well as an improved level of transmission performance obtained by lowering transmission errors.

3.1.1.3.3   Operation

Bluetooth operates at 2.45 GHz in the 2.4-GHz industrial, scientific, and medical (ISM) band allocated for unlicensed transmission. This frequency band is divided into 79 channels, each 1 MHz wide, that a device will randomly hop through by changing channels up to 1600 times per second. This frequency band is also used by garage door openers, baby monitors, cordless phones, and most wireless local area networks (LANs). One of the ways Bluetooth devices avoid interfering with other devices using the same general frequency is through employing very weak signals, which limit the range of Bluetooth devices. A second method used by Bluetooth devices to minimize interference is its spread-spectrum frequency-hopping scheme. By hopping from one frequency to another using up to 79 randomly selected frequencies within a designated range of frequencies that can change 1600 times per second, the possibility of interference with other devices is minimized.

3.1.1.3.3.1   Network Formation — When two or more Bluetooth-capable devices come within range of one another, a handshaking process occurs to determine if they have data to share or if one device needs to control another. This handshaking occurs automatically, and they then form a PAN, also referred to as a piconet due to its relatively small area of coverage. Once a piconet is established, members of the network randomly hop frequencies using the same random frequency selection process so that they can both communicate with other members of the piconet and avoid interfering with other piconets that may be operating in the general area. Each piconet can have up to eight devices, with each device transmitting a set of information at the request of another device. Such information can include the device name, class, list of its services, device features, manufacturer, and Bluetooth version.

3.1.1.3.3.2   Device Addresses — Although each Bluetooth device has a unique 48-bit address, such addresses are generally hidden and replaced by "friendly" names, such as Nokia 3260, which is a cell phone. Such friendly names can usually be set or reset to a default by the user. To illustrate the use of Bluetooth, let us assume you are using a Bluetooth headset with a Bluetooth-capable cell phone. Both devices would form a piconet to talk to the other device. Both the headset and cell phone have distinct 48-bit addresses. When the headset is turned on, it transmits radio signals asking for a response from any units with an address within a particular range assigned to cell phones. The cell phone responds and forms a piconet consisting of two devices. Now, if either device receives a signal from another Bluetooth device, it will ignore the signal because it is not from within the network.

3.1.1.3.4   Advantages and Disadvantages

There are several advantages and disadvantages associated with the use of Bluetooth within a vehicle. In this section we will briefly discuss those advantages and disadvantages.

Many vehicles in the luxury category now include built-in wiring designed to enable a cell phone to be integrated into the infotainment system. For example, many Lexus and Mercedes vehicles come prewired to enable a cell phone to be connected to the wiring so that an incoming call automatically lowers or turns off the speakers and allows the speakers to be used for the call. Unfortunately, this cabling integration only works with a limited number of cell phones, as most manufacturers use a proprietary interface. To overcome this limitation, many vehicle manufacturers market a Bluetooth adapter that will work with any cell phone that has a Bluetooth transmission capability. Thus, instead of forcing a customer to acquire a new cell phone, the use of a Bluetooth telephone adapter can solve many wiring problems.

Because of the cost associated with many Bluetooth telephone adapters, up to $500, many persons will purchase a Bluetooth headpiece that enables the vehicle operator to perform hands-free cell phone operations when using certain PDA-type smart phones, such as the Palm Treo 700W. With voice recognition software, the driver can say "Dial Fred" and the phone will search its directory and dial the number associated with the person's name. However, because the use of a headset is not integrated into the vehicle's infotainment system, the driver will manually have to turn down or mute any CD or radio usage.

3.1.1.4   Wireless LANs

Today there are five standardized types of wireless LANs. In this section we will briefly describe each standard to obtain an appreciation of their operational capabilities. However, because some persons may not be aware of how the frequency-hopping spread spectrum (FHSS) and direct-sequence spread spectrum (DSSS) operate, we will also discuss this. Once we complete our tour of the five types of wireless LANs, we will examine the advantages and disadvantages associated with their use with respect to inter- and intra-vehicle communications.

3.1.1.4.1   The 802.11 Standard

As a brief review of information presented in the previous chapter, the first wireless LAN standard developed by the Institute of Electrical and Electronics Engineers (IEEE) was 802.11. This standard defined the use of three physical layers for wireless communications: infrared, frequency-hopping spread spectrum (FHSS), and direct-sequence spread spectrum (DSSS). Although vendors developed products that used FHSS and DSSS for wireless LANs during the late 1990s, to the best of this author's knowledge no products were ever developed to follow the IEEE 802.11 infrared communications standard.

3.1.1.4.1.1   FHSS — Under the frequency-hopping spread spectrum (FHSS) a station transmits for a small period of time at one frequency, with the period referred to as dwell time, and then hops to a different frequency to continue communications. The frequency-hopping algorithm is known to each LAN station,

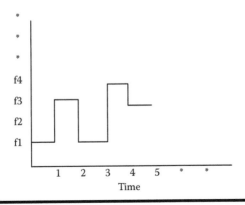

Figure 3.2   Under FHSS communications the carrier frequency is varied in discrete increments based upon a predefined code or algorithm.

enabling each station to adjust its transmitter or receiver according to its mode of operation. One of the more interesting aspects of FHSS is that its origin dates to the actress Hedy Lamarr, who suggested the technique to the U.S. War Department during WWII as a transmission security mechanism.

Figure 3.2 illustrates an example of frequency-hopping spread-spectrum communications. Note that the code or algorithm used to define the manner by which frequencies change can be selected to avoid interference from other non-spread-spectrum communications systems. For example, assume the frequency band f_1 to f_n is used for frequency hopping but f_x, where $f_1 < f_x < f_n$, cannot be used due to its assignment to another use; the algorithm can be altered to preclude the use of f_x.

3.1.1.4.1.2   DSSS — Direct-sequence spread spectrum (DSSS) represents a second transmission technique developed by the military to overcome enemy jamming. Under DSSS, a spreading code is applied to each bit to spread the transmission. At the receiver a majority rule rule is applied. That is, if the spreading code is five bits and the bits received were 10110, because three bits are set the receiver would assume the correct bit is a 1. Under the IEEE 802.11 standard an 11-bit spreading code is employed. The 11-bit spreading code used by DSSS represents a redundant bit pattern that is applied to each information bit to be transmitted. This bit pattern is referred to as a *chip* or *chipping code*, and each bit in the chipping code is module-2 added to the information bit, with the result then transmitted. At the receiver the same chipping code is used to recover the information bit that was transmitted, using the previously described majority rule. That is, the transmitted bits are then module-2 subtracted from the chipping code bits. To illustrate this concept, let us assume for ease of illustration purposes that the chipping code is 5 bits in length instead of 11. Let us focus our attention upon the top portion of Table 3.2, which illustrates the encoding and transmission of four information bits

Table 3.2   DSSS Coding Example

At the Transmitter	
Information bits	1011
Chipping code	10101
Transmitted bits (module 2 addition)	01010 10101 01010 01010
At the Receiver	
Received bits	01010 10101 01010 01010
Chipping code	10101
Reconstructed information bits	11111 00000 11111 11111
Recovered information bit	1011

using a chipping code of 10101. Note that because each information bit is module-2 added to each bit in the chipping code, a total of 20 bits will be transmitted to represent the four information bits. For example, the first (far right) information bit is a binary 1. When module-2 added to the chipping code, the result is the bit sequence 01010, which is transmitted to represent the binary 1 information bit. At the receiver the bit sequence 01010 is modulo-2 subtracted from the chipping code of 10101, resulting in the bit sequence 11111 at the receiver. Because the number of 1 bits exceeds the number of 0 bits, the information bit is considered to be a binary 1. The next information bit is also a binary 1, and the process just described occurs again. Next, the third information bit, which is a binary 0, is modulo-2 added to the chipping code, resulting in the bit sequence 10101, which is transmitted. At the receiver the bit sequence 10101 is received and module-2 subtracted from the chipping code of 10101, resulting in the bit sequence 00000. Because the number of 0 bits exceeds the number of 1 bits, the information bit is assumed to be 0.

Note that our example using a five-bit chipping code enables two bit errors per transmitted bit to have no effect upon the recovered information bit. This results from the fact that the use of a five-bit chipping code and the majority rule enables two bit errors to have no effect upon the correct recovery of each information bit. When the 11-bit chipping code is used, up to 5 bits in the 11-bit transmitted sequence used to represent each information bit can be received in error and have no effect upon the recovered information bit.

3.1.1.4.1.3   Frequency Utilization — Under the basic IEEE 802.11 wireless LAN series of standards the use of the 2.4-GHz frequency band covers frequencies from 2.4 to 2.483 GHz, resulting in 83 MHz of bandwidth being available for use. This frequency band is also referred to as the unlicensed ISM (industrial, scientific, and medical) band. In the United States the Federal Communications Commission (FCC) regulates radiated antenna power to 1 W. In Europe the radiated power level is limited to 10 mW per 1 MHz, while in Japan it is limited to 10 mW.

Table 3.3 FHSS Variations by Geographic Area

Geographic Area	Frequency (MHz)	Hopping Channels	Maximum Transmit Power
North America	2400–2483.5	At least 75; 79 used	1 W
Europe	2400–2483.5	At least 20; 79 used	100 mW
Japan	2471–2497	At least 10; 23 used	10 mW/MHz

In addition to regulatory constraints on power there are differences in the allocation of certain frequencies within the 2.4-GHz band between countries that render equipment operable in one area problematic when used in a different geographic area. Table 3.3 provides a comparison of FHSS by frequency band, hopping channels, and maximum transmit power for three geographic areas.

Similar to FHSS, the frequency allocations for DSSS can vary based on different regulatory agencies. Table 3.4 lists the allowed DSSS center frequencies and corresponding channel numbers for North America, Europe, and Japan.

3.1.1.4.1.4   Utilization —The initial use of 802.11 wireless LANs was limited to their relatively low data rate. This was because each of the three physical layers was only defined for operations at 1 and 2 Mbps. Recognizing the need for a higher data transmission rate resulted in the IEEE initially developing two extensions to the basic 802.11 standard. Those extensions are known as the 802.11a and 802.11b standards.

Table 3.4 DSSS Frequency Utilization

Channel	Frequencies in MHz		
	North America	Europe	Japan
1	2412	n/a	n/a
2	2417	n/a	n/a
3	2422	2422	n/a
4	2427	2427	n/a
5	2432	2432	n/a
6	2437	2437	n/a
7	2442	2442	n/a
8	2447	2447	n/a
9	2452	2452	n/a
10	2457	2457	n/a
11	2462	2462	n/a
12	n/a	n/a	2484

3.1.1.4.2   The 802.11a Standard

The IEEE 802.11a standard defined a series of new modulation methods that enable data transmission rates up to 54 Mbps. The higher data rates are obtained by the use of orthogonal frequency division multiplexing (OFDM), a technique in which the frequency band is divided into subchannels that are individually modulated. A total of 52 subcarriers are used, with 48 used to transmit data, while the remaining 4 are pilot subcarriers that have a carrier separation of 0.3125 MHz (20 MHz/64). Each of the subcarriers can be modulated using binary phase shift keying (BPSK), quadrature phase shift keying (QPSK), 16-position quadrature amplitude modulation (16-QAM), or 64-position quadrature amplitude modulation (64-QAM). In actuality, 802.11a specifies three frequency ranges: 5.15 to 5.35 GHz, 5.47 to 5.725 GHz, and 5.725 to 5.875 GHz. The first two frequency bands are for use indoors, while the last is for outdoor use.

Table 3.5 illustrates the operating frequencies and maximum power of the 802.11a standard for use in the United States. Note that the lower and middle bands provide spacing for eight channels in a total bandwidth of 200 MHz while the upper band provides for four channels in the 100-MHz bandwidth.

The IEEE 802.11a standard defines operations in the 5-GHz frequency band. Because high frequencies attenuate more rapidly than low frequencies, this results in 802.11a wireless LAN stations having a shorter range than stations operating in the 2.4-GHz band. This in turn requires an organization to deploy more access

Table 3.5   802.11a Frequency and Power

Band	Channel Numbers	Frequency (MHz)	Maximum Output Power
5.15–5.25 GHz	36	5180	
	40	5200	40 mW
	44	5220	(2.5 mW/MHz)
	48	5240	
5.25–5.35 GHz	52	5260	
	56	5280	200 mW
	60	5300	(12.5 mW/MHz)
	64	5320	
5.725–5.825 GHz	149	5320	
	153	5745	800 mW
	157	5765	(50 mW/MHz)
	161	5785	
		5805	

points to obtain a similar geographical area of coverage than would be required via the use of access points operating in the 2.4-GHz band.

3.1.1.4.3 The 802.11b Standard

The second extension to the basic IEEE 802.11 standard is the 802.11b standard. Under the IEEE802.11b standard DSSS was used with a new modulation method to provide a data transfer rate of 11 and 5.5 Mbps. The 802.11b standard uses complementary code keying (CCK) as its modulation standard, which can be considered to represent a variation of CDMA. Through an adaptive rate selection process where signal quality is measured, fallback data rates of 5.5, 2, and 1 Mbps can be automatically selected. To provide compatibility with 802.11 DSSS equipment, the IEEE802.11b standard specifies the use of the 2.4-GHz frequency band.

3.1.1.4.4 The 802.11g Standard

As previously noted in Chapter 2, a comparison of the IEEE802.11a and 802.11b extensions to the the 802.11 standard indicates advantages and disadvantages associated with each. Although the 802.11a standard provides a higher data transfer rate, its use of the 5-GHz frequency band results in a shorter transmission distance. Similarly, in a reverse manner, the IEEE802.11b standard provides a greater transmission distance but lower data rate than obtainable from the use of 802.11a-compatible equipment. By combining the modulation method used in the 802.11a standard with the frequency band employed by the 802.11b standard, the IEEE provided a mechanism to extend both the data rate and transmission range of wireless LANs, resulting in the 802.11g standard. To provide backward compatibility with the large base of 802.11b equipment, the 802.11g standard also supports DSSS operations at 11, 5.5, 2, and 1 Mbps. Thus, IEEE802.11g can be considered to represent a dual standard because it provides 802.11b compatibility.

An 802.11g device uses OFDM for data rates of 6, 9, 12, 18, 24, 36, 48, and 54 Mbps. When operating as an 802.11b device it uses CCK for 5.5 and 11 Mbps and DBPSK/DQPSK plus DSSS for operations at 1 and 2 Mbps. When operating as an 802.11b device 802.11g provides a slightly greater range; however, when operating as an 802.11g device at 54 Mbps its range is significantly shorter than that of an 802.11b device.

The 802.11g standard was promulgated in June 2003. By late 2006 this standard was the most popular of all IEEE wireless LAN standards, and products that were formerly available as dual band/dual mode became dual-band/tri-mode adapter cards and access points, as most 802.11g products supported 802.11g, 802.11b, and 802.11a standards.

3.1.1.4.5 The 802n Standard

The newest wireless LAN standard builds upon previous 802.11 standards by adding multiple-input, multiple output (MIMO) capability to provide a theoretical maximum data rate of 540 Mbps. Referred to as 802n, this standard is expected to be finalized by either late 2007 or early 2008.

The key to the ability of the 802.11n standard to provide higher data transfer rates is in its support for multiple transmitter and receiver antennas. Through the use of spatial division multiplexing, several independent data streams can be simultaneously transmitted within one spectral channel of bandwidth. When combined with multiple antennas, this enables multiple spatial division multiplexed data streams to flow from transmitter to receiver at the same time, considerably boosting data throughput.

Currently, about a dozen vendors offer MIMO products that are compatible with the current draft 802.11n standard. Many of these products provide a data rate of approximately 200 Mbps at distances equivalent or slightly exceeding those obtainable by 802.11g products. Table 3.6 provides a summary of IEEE 802.11 wireless transmission standards with respect to their maximum data rate.

3.2 MANET

In this section we will turn our attention to the Mobile Ad Hoc Network (MANET) as it is expected that this type of network will provide the mechanism for most inter-vehicle communications. Because MANET is a mesh network, we will first review the characteristics of this type of network. To differentiate the evolving IEEE 802.11s standard for wireless mesh networks, we will next briefly discuss this standard. Once this is accomplished, we will then turn our attention to the primary focus of this section.

Table 3.6 IEEE 802.11 Transmission Standards

Standard	Maximum Data Rate
802.11	2 Mbps
802.11a	54 Mbps
802.11b	11 Mbps
802.11g	54 Mbps
802.11n	200+ Mbps

3.2.1   Mesh Networking

Mesh networking represents a technology that allows different types of data, to include digitized voice, to be routed between nodes that join and disengage from a network structure on a dynamic basis.

3.2.1.1   Network Formation

Figure 3.3 illustrates a mesh network formed by five nodes, with an additional node in the process of joining the network while another node is in the process of leaving the network. By allowing for continuous connections and disengagements on an ad hoc basis, the mesh network automatically configures itself and notes primary and alternate routes between nodes, allowing data to flow from one node to another around blocked paths by hopping from one node to another until the destination is reached. Here the term *ad hoc* can be considered to be synonymous with the term *on the fly* as nodes randomly join and leave the network.

In examining Figure 3.3 note that each node needs to keep track of the other nodes in the network so that data can be sent to other nodes when required. Also note that when a node joins or leaves a network the presence or absence of the node must be conveyed to other nodes.

3.2.1.2   Area of Coverage

A basic mesh network can cover a geographic area ranging in size from a group of buildings on a campus to a large city such as Philadelphia, which is probably the first large North American city to provide WiFi services to its inhabitants via a mesh network. When used to cover a predefined geographic area, the mesh network has a series of fixed access points, while the nodes representing wireless stations move

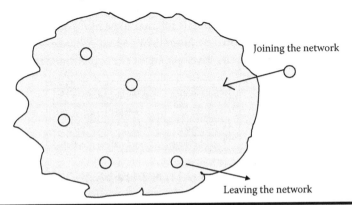

Figure 3.3   A mesh network provides a method to route data between nodes.

about the coverage area, but for the most part are relatively fixed with respect to their location. A second type of mesh network involves stations that are almost always moving and stop for short periods. This type of mesh network is referred to as a Mobile Ad Hoc Network (MANET) and represents the type of network that would more closely satisfy the requirements of vehicles in a mobile environment.

The key difference between a MANET, or what some persons refer to as a Vehicular Ad Hoc Network (VANET), and a conventional wireless mesh network is in two areas. First, a MANET has no fixed points, while a conventional wireless mesh network does in the form of access points. The second difference is in communications capability, because in a MANET every node is usually able to communicate with every other node, while in a mesh network every node needs to communicate with an access point. Now that we have a general appreciation for the differences between wireless mesh networks and MANETs, let us turn our attention to the IEEE evolving standard being developed for the former prior to discussing the latter in detail.

3.2.1.3   The IEEE 802.11s Standard

Because IEEE 802.11s is being developed to facilitate mesh networking, it represents an evolving standard that overcomes some of the same problems associated with MANET. Thus, by obtaining an overview of this evolving standard we can note the major functions that need to be performed to support a self-configuring multi-hop topology.

The IEEE802.11s represents an evolving standard that is expected to be promulgated during 2008. The goal of this evolving standard is to provide an architecture and protocol for the automatic configuration of paths between access points over multi-hop nodes in a wireless environment. Initially, a call for proposals in June 2005 resulted in 15 submissions that were pared down to 4 by September 2005. In January 2006 two competing proposals referred to as SEE Mesh, led by Intel and Firetide, and Wi-Mesh, proposed by Nortel's Wi-Mesh Alliance, were merged and used as the starting point for the development of the actual standard.

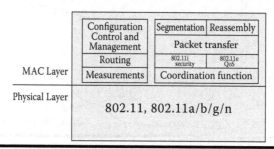

Figure 3.4   Wireless LAN architecture.

3.2.1.3.1   Components

Figure 3.4 illustrates the major components that will, in this author's opinion, form the basis for the 802.11s architecture. Some of these components, such as the physical layer for the series of IEEE802.11 transmission methods (802.11a/b/g/n), represent existing standards that govern how data is transmitted. Other components, such as configuration, control, and management (CCM), define how nodes are configured, controlled, and managed, to include routing and measurements required to update routing tables.

Under the IEEE 802.11 standard there are three different distributed coordination functions (DCFs) that can be used to provide access to the medium. Essentially, the DCFs check to determine if the radio frequency is available prior to transmission as a mechanism to avoid contention. If the link is busy, the station with data to transmit will wait a random back-off time, and if the link is still busy, each time the station listens to the medium the contention window will double up to its maximum value.

If the wireless system supports quality of service (QoS), a new coordination function — the hybrid coordination function (HCF) — is used. HCF has two modes of operation known as Enhanced Distributed Channel Access (EDCA) and HCF-Controlled Channel Access (HCCA). EDCA is a contention-based channel access function that operates concurrently with HCF-controlled channel access based on a polling mechanism. EDCA supports prioritized traffic, while HCCA supports parameterized traffic.

Moving up the architecture diagram shown in Figure 3.4, measurements include paths between nodes as well as the entry and exit of nodes from an ad hoc network. This information is used for routing packets from source to destination. In addition, the CCM management process uses measurements for topology discovery. Other functions performed by CCM include channel allocation, path selection, and packet forwarding.

3.2.1.3.2   Terminology

The introduction of wireless mesh networking resulted in several new terms being added to the wireless literature. First, a *mesh point* represents a node that has the ability to support mesh services. If a mesh point also supports access point services, it is referred to as a *mesh access point*, while a mesh point that provides a connection to a wired network is referred to as a *mesh portal*.

3.2.1.3.3   Routing

In a mesh network a mesh point must be able to discover its peers and associate with them. In addition, the mesh point needs to be able to select an optimum path through the mesh network to forward frames to their destination. Under the evolving 802.11s standard a path selection protocol referred to as Hybrid Wireless Mesh (HWM) is being considered. To enhance interoperability, vendors can also use their own protocols for path selection.

3.2.1.3.4   Summary

It is important to note that the 802.11s evolving standard is being developed as a single-radio, shared mesh extension for indoor access points. This is a very different type of application from the large outdoor wireless infrastructure used for enabling small and large cities to provide citizens with Internet access. In addition, it will also differ from ad hoc mobile networks that will more than likely be used by vehicles. Thus, in concluding this chapter we will focus our attention on MANET.

3.2.1.4   Understanding MANET

The effort in developing a series of specifications for MANET dates to 2001, when the MANET Working Group (WG), part of the Internet Engineering Task Force (IETF), published a draft version of the Ad Hoc On-Demand Distance Vector (AODV) Routing protocol. In July 2003 the latest draft of AODV was published as RFC 3561. To date, the MANET WG has developed six Internet draft documents and five RFCs.

3.2.1.4.1   Working Group Goals

The goal of the MANET WG is to "standardize Internet Protocol (IP) routing protocol functionality suitable for wireless routing applications within both static and dynamic topologies with increased dynamics due to node motion or other factors." The MANET WG intends to support both IPv4 and IPv6 and plans to develop a forwarding protocol that will efficiently flood packets to all participating MANET nodes.

3.2.1.4.2   Routing Protocols

Over the years literally hundreds of routing protocols have been developed, with most protocols designed to perform specific tasks. In this section we will examine some distinct characteristics of routing protocols.

3.2.1.4.2.1   Characteristics — There are two major characteristics that can be used to distinguish routing protocols: the manner by which they gather routing data and how they route data toward their destination.

Routing data generation — Concerning the generation of routing data, a protocol either is table driven (proactive) and has complete up-to-date routing information or only gathers routing information when needed (on demand). If the ad hoc routing protocol is table driven, a certain overhead is associated with its use, which significantly increases as the number of nodes and their mobility increase. In comparison, the use of on-demand routing information minimizes overhead but introduces a routing delay, as a node must gather information to route data through the network. Thus, we can say that an on-demand routing protocol introduces delays that grow as the size of the network increases. An on-demand routing protocol is also referred to as a reactive protocol.

Routing data to a destination — A second characteristic of routing protocols concerns the manner by which they route data toward the destination. In some protocols destinations are based upon each node having a map of the network that shows which nodes are connected to other nodes (link state), while other routing protocols have each node share its routing table with its neighbors (distance vector). If a link state protocol is used, each node knows all details of the routes and includes complete information about the routing path in every transmitted packet, which increases overhead. If a distance vector protocol is used, each node only knows about its neighbors and the next hop toward the packet destination, which minimizes node overhead. When we consider the use of inter-vehicle communications via a WiFi infrastructure, we can envision a large network with high mobility. Thus, a reactive or on-demand distance vector protocol would appear to be a good fit for routing within a MANET. With this in mind, let us examine how AODV operates.

3.2.1.4.3   AODV Operations

AODV, as its name implies, is a routing protocol categorized as a distance vector protocol where every node knows its neighbors and the cost to reach them. Each node in the network maintains its own routing table, which results in each node discovering and eventually storing the distance and next hop information for all nodes in the network. In the event a node is not reachable, the distance is set to infinity.

3.2.1.4.3.1   Table Updates — On a periodic basis each node transmits its routing table to its neighbor nodes. Those nodes examine the routing tables of their neighbors to determine if a route to another node using the neighbor as the next hop is useful. If so, this information is added to the node's routing table. Because each node has its own routing table, this routing protocol has a relatively short delay. AODV supports unicast, broadcast, and multicast transmission without requiring any additional protocol support. If a link between nodes breaks, a misbehavior called count-to-infinity can occur. Count-to-infinity and loop problems are solved using sequence numbers in the exchange of data and assigning applicable

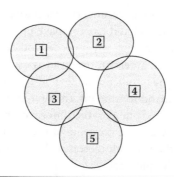

Figure 3.5   A five-node AODV network.

cost. To illustrate the manner by which AODV operates, we need a network. Thus, let us assume we have a five-node network whose nodes are labeled 1 through 5, as illustrated in Figure 3.5

3.2.1.4.3.2   Using HELLO Messages — The circles in Figure 3.5 indicate the transmission range for each node. For example, node 1 can communicate with nodes 2 and 3 but not directly with nodes 4 and 5. Now let us assume that node 1 wants to transmit a message to node 5. Unfortunately, node 1 cannot communicate directly with node 5 or, for that matter, with node 4, as they are not within range. However, nodes can communicate directly with neighboring nodes. To do so, nodes keep track of their neighbors by listening for a HELLO message that each node broadcasts at set intervals. In actuality, a HELLO message is a Route Reply (RREP) message with a time-to-live (TTL) value of 1. Later in this section we will examine the format of the Route Reply message.

Based upon the exchange of HELLO messages, each node constructs a routing table that indicates the nodes in the network, the next hop node, and the hop count to the next hop. Table 3.7 illustrates the series of routing tables that are eventually created due to the exchange of HELLO messages.

To interpret Table 3.7, let us focus our attention upon node 1. Node 1 can reach nodes 2 and 3 directly, so the next hops for node 1 are nodes 2 and 3, each with a hop count of 1. To go from node 1 to node 4 requires first going to node 2 and then to node 4, for a hop count of 2. Similarly, to transmit data from node 1 to node 5 requires node 1 to first transmit to node 3, which then transmits to node 5. Thus, the hop count is 2.

3.2.1.4.3.3   Route Request Message — When one node needs to transmit a message to another node that is not its neighbor, it broadcasts a Route Request (RREQ) message. Figure 3.6 illustrates the format of the RREQ message.

Table 3.7 Routing Tables Created for the Five-Node Network

Node 1

Node	Next Hop	Hop Count
2	2	1
3	3	1
4	2	2
5	3	2

Node 2

Node	Next Hop	Hop Count
1	1	1
3	1	2
4	4	1
5	4	2

Node 3

Node	Next Hop	Hop Count
1	1	1
2	1	2
4	5	2
5	5	1

Node 4

Node	Next Hop	Hop Count
1	2	2
2	2	1
3	5	2
5	5	1

Node 5

Node	Next Hop	Hop Count
1	3	2
2	4	2
3	3	1
4	4	1

Type	J	R	G	D	U	Reserved	Hop Count
RREQ ID							
Destination IP Address							
Destination Sequence Number							
Originator Sequence Number							
Originator Sequence Number							

Figure 3.6   RREQ message format. J = join flag (reserved for multicast); R = repair flag (reserved for multicast); G = gratuitous flag; D = destination-only flag; U = unknown sequence number flag; hop count = number of hops from originator to node handling the request; RREQID = a sequence number uniquely identifying the RREQ when used in conjunction with the originator's IP address; destination IP address = IP address of destination for which a route is desired; destination sequence number = the latest sequence number received by the originator for any route toward the destination; originating IP address = the IP address of the node that originated the Route Request; originator sequence number = the current sequence number to be used in the route entry pointing toward the originator of the Route Request.

As indicated in Figure 3.6, the RREQ message contains the source and destination IP addresses and a pair of sequence numbers that serve as unique identifiers when examined in conjunction with source and destination IP addresses. To illustrate the use of RREQ messages, let us assume that node 1 needs to send a message to node 5. Node 1's neighbors are nodes 2 and 3. Because node 1 cannot directly communicate with node 5, node 1 transmits an RREQ that is heard by nodes 2 and 3.

3.2.1.4.3.4   Route Reply Message — When node 1's neighbors receive the RREQ message, they have two choices. If they know a route to the destination or they are the destination, they can transmit a Route Reply (RREP) message back to node 1. If neither situation occurs, they will rebroadcast the RREQ message to their neighbors. Their neighbors will then continue the rebroadcasting until the life span is reached. Note that if node 1 does not receive a reply within a predefined time, it will rebroadcast the RREQ message, using a longer life span and a new RREQ ID number. Concerning the sequence number, all nodes use that number in their RREQ as a mechanism to ensure that they do not rebroadcast an RREQ.

Returning to Figure 3.6 and Table 3.7, node 3 has a route to node 5 and responds to the RREQ transmitted by node 1 by transmitting an RREP to node 3, which then sends it to node 1. Figure 3.7 illustrates the format of the Route Reply (RREP) message.

Type	R	A	Reserved	Prefix Size	Hop Count
Destination IP Address					
Destination Sequence Number					
Originator IP Address					
Lifetime					

Figure 3.7 Route Reply (RREP) message format. Type = 2; R = repair flag (used for multicast); A = acknowledgment required; prefix size = specifies that the indicated next hop may be used for any nodes with the same routing prefix; hop count = number of hops from the originator to destination; lifetime = time in milliseconds for which nodes receiving the RREP consider the route to be valid; destination IP address, originator IP address, and destination sequence = numbers are the same as defined for an RREQ message.

3.2.1.4.3.5   Sequence Numbers — Sequence numbers in an RREQ message function as a time stamp. That is, they allow nodes to determine how "fresh" their information is concerning other nodes. Each time a node transmits any type of message it increases its own sequence number. Each node records the sequence number of all other nodes it talks to, with a higher sequence number indicating a fresher route. Thus, the data shown in Table 3.7, while representing the routing between hops in Figure 3.5, is actually incomplete; as in an AODV environment, each routing table would include a sequence number column that would enable nodes to update their entries as the topology changes.

3.2.1.4.3.6   Error Messages — AODV includes a Route Error (RERR) message, which provides the protocol with the ability to adjust routes when nodes move around the network. There are three conditions under which a node will broadcast an RERR message to its neighbors. When any one of these three conditions occurs the node receiving the RERR removes all routes that contain the bad nodes. The three conditions are:

1. When a node receives a data packet that it is supposed to forward but does not have a route to the destination.
2. A node receives an RERR message that causes at least one of its routes to become invalid. When this occurs the node transmits a new RERR with all of the new nodes that cannot be reached.
3. A node detects that it cannot communicate with one of its neighbors. When this situation occurs the node examines its route table for the route that uses the neighbor for a next hop and marks the entry invalid, after which it transmits an RERR message with the neighbor and the invalid routes.

3.2.1.4.3.7   Unicast Routing — AODV uses three control messages to support unicast routing: RREQ, RREP, and RERR. When a node has data to transmit to another node for which no route is available, it will broadcast an RREQ (Route Request) to find one. As previously shown in Figure 3.6, an RREQ message includes a unique identifier, the source and destination IP addresses, a sequence number, a hop count that is initially set to zero, and flag fields primarily used for multicast routing.

When a node receives an RREQ that it has not previously seen, it sets up a reverse route to the originator. If the receiving node does not know the route to the destination, it rebroadcasts the updated RREQ with an incremented hop count. If the receiving node knows the route to the destination, it simply creates and transmits an RREP (Route Reply) message. The RREP will be unicasted to the originating node to take advantage of the reverse route just established. The RREP, which was shown in Figure 3.7, contains source and destination IP addresses, a sequence number, a time-to-live value in ms, a hop count, and a prefix, with the latter used for subnets and with certain flag bits. When a node receives an RREP it will check to determine if the sequence number in the message is higher than the one in its routing table and if the hop count in the RREP message is lower than the one in its routing table. If neither situation exists, the originator discards the RREP message. Otherwise, a higher sequence number in the message indicates the data is fresher than that stored in the node's routing table and the node updates its routing table. If the node is not the destination, it will rebroadcast the RREP to its neighbor nodes to include the node from which it received the RREP.

To illustrate the route establishment process under AODV, let us begin with another five-node network, which is illustrated in Figure 3.8. Suppose node 1 again needs to transmit data to node 5. Assuming node 1 does not have an entry in its routing table for node 5, it broadcasts an RREQ to its neighbors. This is indicated by the two arrows shown in Figure 3.8.

When nodes 2 and 3 receive the RREQ message, they rebroadcast the message to nodes 5 and 4, respectively. In addition, nodes 2 and 3 establish a reverse route to node 1 by transmitting RREP messages. Figure 3.9 illustrates this action.

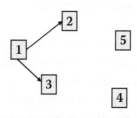

Figure 3.8   Node 1, needing to transmit a packet to node 5, first queries its neighbors.

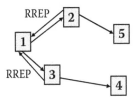

Figure 3.9   Intermediate nodes 2 and 3 rebroadcast RREQs and establish reverse routes to node 1.

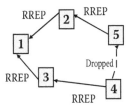

Figure 3.10   The duplicate RREQ from node 4 is dropped by node 5. Node 5 transmits an RREP to node 2 that flows to node 1.

When the RREQ messages reach nodes 4 and 5, they establish reverse routes to nodes 3 and 2. In addition, because the distance from nodes 1 to 2 to 5 is two hops while nodes 1 to 3 to 4 to 5 is three hops, the shortest hop count from node 1 to 5 is via node 2. Thus, node 5 transmits a unicast RREP to node 2 and drops the duplicate RREQ received via node 4. This is shown in Figure 3.10.

Upon receipt of the RREP at node 1 it establishes a route to node 5. This is shown in Figure 3.11.

3.2.1.4.3.8   Summary — AODV or a modification of this protocol is well suited for vehicle ad hoc networking. Through the use of periodic HELLO messages, neighbors are tracked and each node only keeps track of the next hop for

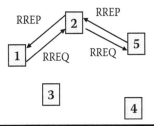

Figure 3.11   Node 1 establishes a route to node 5 while unused reverse routes expire.

a route instead of the entire route, which minimizes overhead. Because AODV only discovers routes as they are needed, transmission is minimized while the use of unique sequence numbers permits the freshness of data to be noted and table entries to be updated with a minimal amount of processing, in which a node only needs to compare sequence numbers in a table entry to those in a message.

Chapter 4

The Local Interconnect Network

In this chapter we commence our examination of the manner by which intra-vehicle communications occurs by focusing our attention upon the Local Interconnect Network (LIN). Beginning with an overview of the evolution and technology associated with LIN, we will examine its key features, how it is used in automobiles and other vehicles, the role of the LIN consortium, and its use as a bridge into a Controller Area Network (CAN), a more robust and costly network introduced by Bosch for in-vehicle networks in 1986.

4.1 Overview

The Local Interconnect Network (LIN) represents a relatively recent low-cost serial communications system developed to provide communications between electronic systems and sensors in vehicles. The first LIN specification dates to 1999, with a major revision, referred to as LIN 2.0, published by the LIN Consortium in September 2003.

4.1.1 Founding Members

The founding members of the LIN Consortium include such European car manufacturers as BMW, Daimler Chrysler, Volkswagen Audi Group, and Volvo Car Corporation. Although initially considered a European specification, the Society of Automotive Engineers (SAE) issued its J2602 recommended practice titled "LIN Network for Vehicle Applications," with representation from Ford and General Motors. The major difference between LIN 2.0 and the SAE J2602 is in the areas of data rate and error handling. Under LIN 2.0 a serial data rate up to 20 kbps is supported. In comparison, under the SAE J2602 specification the data rate is limited to 10.4 kbps.

4.1.2   Goal

LIN represents an open standard that provides a low-cost alternative to the use of the Controller Area Network (CAN), which is also known as SAE J1850. Because LIN represents cost-effective communications for smart sensors and actuators, where the additional capability of CAN is not required, its use provides vehicle designers with the ability to add functionality at a relatively low price. This in turn should result in more reliable vehicles, as LIN can be used to provide a significant level of diagnostics capability at a more affordable price.

In this chapter when we reference LIN we are actually talking about a communications concept for local interconnected networks in a vehicle. This concept covers the definition of physical and data-link layers, how data is transferred, which is the LIN protocol, as well as the interfaces for a series of development tools and application software. As we probe deeper into LIN we will discuss the individual components that collectively represent the concept.

4.1.3   Applications

Due to the relatively low cost of LIN, it becomes possible for this networking technology to support a variety of intra-vehicle applications. Table 4.1 lists some of those applications, categorized by the major vehicle area where the application has its primary control. For example, under the steering wheel control the applications listed involve the regulation of a range of automotive components; however, because those components are controlled via the steering wheel or steering column, they are listed under that category.

In examining the applications listed in Table 4.1 it is reasonable to expect that LIN applications will involve assembly units of a vehicle, such as steering wheels, doors, seats, lighting, and interior climate regulation. Due to the lower cost of LIN in comparison to other wiring control methods, its use makes it possible to introduce more safety-related controls and diagnostics without substantially increasing the cost of a vehicle. In addition, because the use of LIN replaces commonly used analog signaling with digital signaling, the accuracy and self-diagnosis of equipment is improved. Now that we have an appreciation for some of the advantages of the LIN specification and its actual and potential use, let us turn our attention to the latest version of the LIN specification, version 2.0.

Table 4.1   Typical LIN Applications

Steering wheel/steering column
Telescopic control
Cruise control
Wiper control
Turn signal control
Climate control[a]
Radio/CD control[a]
Navigation control[a]
Telephone control[a]
Seat
Forward/backward horizontal position motor
Vertical position motor
Occupancy sensor
Door
Mirror control switch
Window lift
Door lock
Child lock
Seat control switch
Roof
Sun roof control
Light control (light sensor) for interior light
Windshield
Rain sensor
Engine
Various sensors
Various small motors
Climate
Control panel

[a] May be an option or located in another area of the vehicle.

4.2   The LIN Specification

Previously we noted that the LIN specification represents a low-cost serial communications system, but did not actually describe how it operates. In this section we will focus our attention on the latest version of LIN, the LIN 2.0 specification, to include examining how data is transmitted and the reasoning behind developing what many persons might consider to represent a networking technology that has a relatively low data rate transfer capability in an era where we are accustomed to technologies that provide increasingly high data rates.

4.2.1 The Specification Components

The LIN specification is similar to a troika, as it consists of three main parts. Those parts are the LIN Protocol Specification, which describes the physical and data-link layers; the LIN Configuration Language, which describes the format of the LIN configuration file that is used to configure the network; and the LIN Application Programming Interface (API), which describes the interface between the network and the application program. In addition, a pair of tools is available for developers and technicians that expands upon the LIN Configuration Language. Those tools include a network configuration generator and a bus analyzer/emulator, with the latter also referred to as the LIN spector.

Although LIN represents a serial communications interface, the actual specification covers both hardware and software, to include tools that simplify the development and testing of applications. Figure 4.1 illustrates in a block diagram the major components of the LIN specification.

4.2.1.1 The Physical Layer

As a method to minimize cost, LIN uses a single conductor wire to form a 12-V bus that uses the vehicle's body to serve as a common ground. Communications is based upon the serial communications interface (SCI) universal asynchronous

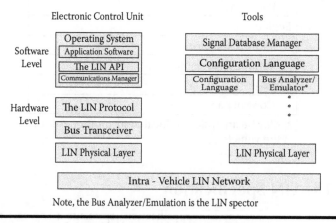

Note, the Bus Analyzer/Emulation is the LIN spector

Figure 4.1   The major components of the LIN specification.

receiver-transmitter (UART) data format, with a clock synchronization used for nodes without a stabilized time base. The maximum data rate is 20 kbps while the maximum transmission distance is 40 m.

4.2.1.2   Master–Slave Relationship

LIN can be considered to represent a broadcast serial network consisting of 1 master and up to 16 slave nodes. Because transmission occurs on a common bus, data reaches all nodes attached to the bus; hence, it becomes a broadcast network. Because no collision detection mechanism exists, all messages are initiated by the master node, with one or more slave nodes responding to a predefined message identifier in the broadcast. Commonly the master node is a moderately powerful microcontroller while slaves are either dedicated application-specific integrated circuits (ASICs) or less powerful microcontrollers.

In the modern automobile several LINs, each with up to 16 nodes, may be in use. Each of the LINs are then interconnected via their master nodes to a more expensive CAN upper layer network, which enables LIN-based information to be forwarded to a central location for projecting warning information on a console display, performing diagnostic testing, and enabling other centralized tasks to be performed.

4.2.1.3   Interference

The 20-kbps maximum data rate can be exceeded; however, doing so can result in electromagnetic interference. Thus, the data rate is kept at or below 20 kbps as a mechanism to minimize electromagnetic interference (EMI). Through self-synchronization, neither crystals nor ceramic resonators are required in slave nodes that are controlled by a master node in a master–slave relationship. Thus, the ability to avoid the use of crystals or ceramics resonators results in a significant cost reduction.

4.2.1.4   Examining the Master–Slave Relationship

The LIN bus consists of a single master with one or more slaves, up to a maximum of 16. The master uses one or more predefined scheduling tables to commence transmission and reception of data on the LIN bus. The scheduling tables contain as a minimum the relative timing with respect to when a message transmission is initiated.

Figure 4.2 illustrates an example of the master–slave relationship, where a master is shown controlling four slaves. Each master node contains a master task and a slave task, while all slaves contain only a slave task.

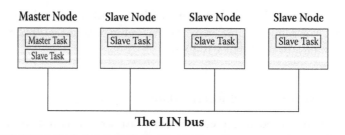

The LIN bus

Figure 4.2   LIN master–slave relationships.

4.2.1.4.1   The Master Task

The master task determines when and what frames should be transferred onto the bus. In comparison, the slave task provides the data transported by each frame.

The master task has control over the entire bus and the protocol, determining which messages are to be transferred over the bus and when the transfer should occur. In addition, the master task can perform error handling. To accomplish the preceding, the master task will transmit a SYNC BREAK to mark the beginning of a message frame; a SYNC byte, which represents a predefined pattern for determination of the time between two rising edges of a pulse; and a message identifier. The last includes information about the sender and one or more receivers, the purpose, and the length of the data field. Other functions performed by the master task include monitoring of data and check bytes and receiving a wake-up BREAK from slave nodes when the bus is inactive and a slave requests some action.

4.2.1.4.2   The Slave Task

In comparison to the master task, the slave task is relatively simple. The slave task either provides a response to the master task or ignores a master task if a requested action by the master is directed to other slaves.

4.2.1.5   Data-Link Layer

In the International Standards Organization (ISO) Open Systems Interconnection (OSI) model, the data-link layer is that above the physical layer. In the wonderful world of the LIN, the data-link layer corresponds to the manner by which frames are transported over the bus. Thus, let us turn our attention to the LIN frame.

4.2.1.5.1   The LIN Frame

Each LIN frame consists of a header and a response. The header is transmitted by the LIN master (master task), while the response is returned by one LIN slave (slave

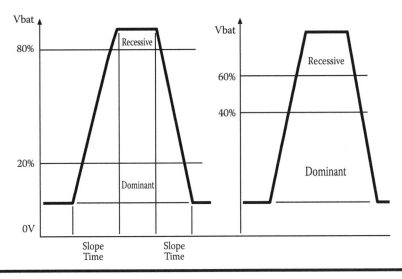

Figure 4.3   LIN bus data.

task). Data is transmitted serially as 8 data bits, with 1 start and 1 stop bit with no parity, resulting in 10 bits transmitted per byte. Data on the bus is divided into recessive (logical high) and dominant (logical low), as shown in Figure 4.3.

As indicated in Figure 4.3, at the transmitter the lower-level voltage should be less than 20 percent of the battery voltage, or approximately 1 V. Doing so results in a logic 0. In comparison, the upper-level voltage, which represents a logic 1, should be more than 80 percent of the battery voltage. At the receiver a logic 0 will be less than 60 percent of the battery voltage, while a logic 1 will be more than 60 percent of the battery voltage.

4.2.1.5.1.1   Frame Header — The frame header consists of three parts: a BREAK field, a SYNC pattern field, and an identifier field. In the following sections we will become acquainted with each field.

BREAK Field — The BREAK symbol is used to identify the beginning of a new frame. In addition, this 1-byte field is used to activate all LIN slaves on the bus to listen to the following parts of the header. The BREAK field consists of 1 start bit and a minimum of 13 SYNC BREAK bits. The purpose of the BREAK field, in addition to activating slaves to listen to the following parts of the header, is to ensure that listening nodes with a main clock differing from the set bus signaling rate detect the BREAK as the frame header and not as a data byte with all values set to zero.

Synchronization Field — The SYNC pattern is a standard data byte with a hex value of 55. This byte, as previously mentioned, is used for determining the time between two rising pulse edges.

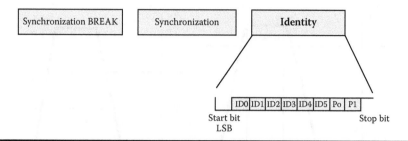

Figure 4.4 Components of the LIN frame header.

Identifier Field — The third field in the LIN frame is the identifier, or ID field. This field defines an action to be performed by one or several of the LIN slaves attached to the bus. If the identifier results in one physical LIN slave sending a response, the identifier is referred to as an Rx identifier. In comparison, if the master slave task sends data to the bus, it is then referred to as the Tx identifier.

The ID field message identifier has a length of six bits. These six bits are coded via tables that provide information about the sender, receivers, purpose of the message, and length of the data field. The data field can be either 2, 4, or 8 bytes, with the length of the data field indicated by the two most significant bits of the ID field. Because of the importance of the message identifier, it is protected through the use of two linked parity bits.

Figure 4.4 illustrates the components of the header portion of a message frame. Note that the header is transmitted by the master node while the response is returned by a slave node. Together, both the header and response make up the message frame.

The six bit positions in the identifier result in 64 possible values, from 0 to 63, that can be set. Values 0 to 59 are used to identify signal-carrying frames, while values 60 and 61 are used to carry diagnostics. Currently, the value of 62 is reserved for user-defined extensions, while the value of 63 is reserved for future product enhancements. Because the identifier value is extremely important, it is protected through the use of two parity bits, P_0 and P_1.

P_0 is computed by XORing, IDO, ID1, ID2, and ID4. That is,

$$P_0 = IDO \text{ XOR } I1 \text{ XOR } ID2 \text{ XOR } ID4$$

The second parity bit, P_1, is computed by first XORing ID1, ID3, ID4, and ID5, after which the result is inverted such that a 0 result becomes a binary 1 and a binary 1 result becomes a binary 0. Thus, the equation for P_1 is

$$P1 = ID1 \text{ XOR } ID3 \text{ XOR } ID4 \text{ XOR } ID5$$

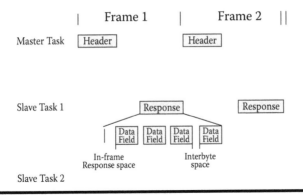

Figure 4.5 Relationship between master task and slave tasks.

4.2.1.5.1.2   Response — The response to a header frame is provided by one of the LIN slave tasks attached to the bus. The slave task receives or transmits data when an applicable message identifier is received. The slave task waits for a SYNC BREAK and then synchronizes on the following SYNC byte. Once synchronization is accomplished, the slave examines the message ID to determine if it should receive data, transmit data, or ignore the message. When transmitting, the slave will send 2, 4, or 8 data bytes as well as a checksum consisting of a single byte of data.

Figure 4.5 illustrates the relationship between the master task and slave task. Remember, under the LIN architecture a frame consists of a header and a response. Thus, the relationship between the master task and slave tasks shown in Figure 4.5 in effect represents the flow of two frames on the LIN bus.

Data — Under the LIN specification two types of data can be transported in a frame: signals or a diagnostic message. A signal is a scalar value or byte array that is packed into the data field of a frame. The position of the signal is at the same location in the data field for all frames that have the same identifier. In comparison, diagnostic messages are carried in frames that have two reserved identifiers. When a diagnostic message is transported, the interpretation of the data depends upon the contents of the data field and the state of the communicating nodes.

A frame can transport from one to eight data bytes, with each data byte transmitted in a byte field. When more than one data byte is transmitted the least significant byte (LSB) is transmitted first while the most significant byte (MSB) is transmitted last. Figure 4.6 illustrates the numbering of data bytes when a frame contains eight such bytes.

Checksum — The last field in a LIN frame is the checksum. The checksum is computed by inverting the summing of all values of the data bytes and subtracting 255 each time the sum is greater or equal to 256. Note that the summing process is not equivalent to a module 255 or module 256 process.

Figure 4.6 Numbering of data bytes in a frame.

Types — There are two types of checksums: classic and enhanced. When the checksum is computed only over the data bytes, it is referred to as a classic checksum. This type of checksum is used for communications with LIN slaves. The second type of checksum is computed over the data bytes and the protected identifier byte. This checksum is referred to as an enhanced checksum and represents the method of communications with LIN 2.0 slaves.

Because there is no error handling routine in a LIN, faulty LIN messages are simply considered unsent and are rejected. Thus, parity and checksum errors result in a message being considered unsent and rejected. However, detected errors are stored in specific slave controllers and can be read by the LIN bus master, which enables error conditions to be noted.

Frame Types — There are two types of frames defined with respect to conditions that invoke frame transmission: unconditional and event triggered.

Unconditional — An unconditional frame always transports signals and their identifier within the range of 0 to 59. This type of frame is always initiated by the master and has a single publisher, although it can have one or multiple subscribers.

When the master task processes a frame allocated to an unconditional frame, the header of this frame is transmitted. The slave task then provides the response. Figure 4.7 illustrates a master controlling two slaves on a LIN bus. First, the master requests a frame from slave 1, which then responds with data. This is followed by the master transmitting an unconditional frame to both slaves, informing slave 2 to respond to slave 1. Thus, slave 2 then transmits a frame to slave 1.

Event Triggered — The second type of frame is event triggered. This frame flows on the LIN bus as a result of the occurrence of an activity. Thus, an event-triggered frame minimizes traffic on the bus because it eliminates or reduces the need for a master to poll slave nodes for seldom occurring events.

An event-triggered frame will use the data field of one or more unconditional frames. Thus, the identifier of an event-triggered frame is the same as that for an

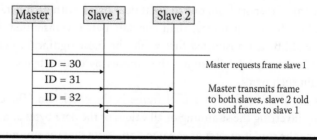

Figure 4.7 Transmitting a sequence of three unconditional frames.

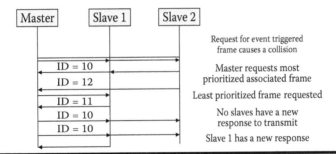

Figure 4.8   An example of an event-triggered frame sequence.

unconditional frame, in the range of 0 to 59. The first data byte in the event-triggered frame that uses the data field of an unconditional frame is equal to the identifier field. Thus, a maximum of seven bytes of data can be included in an event-triggered frame. When more than one unconditional frame is associated with one event-triggered frame, they are all of equal length and use the same checksum type.

When an event trigger action occurs, the result is based upon the response to the frame header. If no slaves respond, the header is ignored. If more than one slave responds, a collision will occur that the master must resolve. To do so, the master will request all associated unconditional slaves prior to again requesting an event-triggered frame. Figure 4.8 illustrates the data flow on a LIN bus when a master's request for an event-triggered frame results in a collision. Note that when a collision occurs the master will first request the most prioritized associated frame.

Other Frame Types — In addition to unconditional and event-triggered frames, LIN supports diagnostic and user-defined frames. Diagnostic frames always convey diagnostic or configuration data and use eight data bytes. A diagnostic frame requested by the master has an identifier of 60, while a slave response diagnostic frame has an identifier of 61. In comparison, a user-defined frame has an identifier of 62.

Operational Example — One example of the use of a triggered event is the monitoring of the child door lock. By using an event-triggered frame to poll the status of the child door lock, the transfer of data on the LIN bus is minimized.

4.2.1.5.1.3   Transmission Events — There are two methods by which frames are transmitted on a LIN bus: a time-triggered approach and a message sequencing approach.

Time Triggered — In the time-triggered approach the message length is known. Thus, the number of bytes to be transmitted is also known, which enables the minimum length to be computed. Because each message has a length budget of 140 percent of its minimum length, the maximum length of a message can be determined. Because the time between the transmission of a header and a slave's

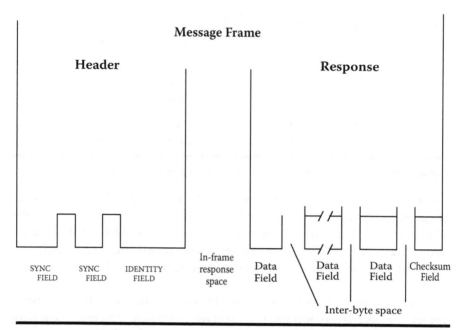

Figure 4.9 Transmitting a frame using the time-triggered approach.

response is separated by an in-frame response space, it becomes possible to predict the flow of the header and response that make up a message frame. Figure 4.9 illustrates a frame transmitted using the time-triggered approach.

Message Sequencing — In a message sequencing approach the message sequence is known and the master uses the scheduling table. The scheduling table denotes the identifiers of each header and the interval between the start of a frame and the start of the following frame. Thus, the master node uses the scheduling table to control both the timing and scheduling of frames transmitted on the bus. An example of message sequencing by the master node using the scheduling table is shown in Figure 4.10. In this example messages A, B, and C are followed by messages B, A, and E.

Through the use of different scheduling tables it becomes relatively easy to alter the sequence that messages flow on the bus. Thus, message sequencing provides a high degree of flexibility, as it can be used to alter the manner by which messages flow on the bus.

Message A	Message B	Message C	Message B	Message A	Message E

Figure 4.10 Using the schedule table.

4.3 Summary

The Local Interconnect Network is a serial bus that enables a master node to control up to 16 slaves. Although the data transfer rate is limited to 20 kbps, the network architecture provides a robust, low-cost mechanism for controlling electronic-based devices distributed throughout a vehicle. With the ability to interface the more expensive CAN, it becomes possible to interconnect individual LINs so that messages can flow to a central location and be displayed for viewing by the vehicle operator. For example, the state of seat belt usage, locking of remote mirrors, control of windows, climate control, and interior lighting could be placed on separate LINs, with the separate networks tied together by each master node interfacing a common CAN.

5.5 Summary

Chapter 5

The Controller Area Network

In the previous chapter we turned our attention to the Local Interconnect Network (LIN). As a brief refresher, LIN represents a low-cost serial communications system that enables a master node to control up to 16 slaves. In addition, it is common to interconnect two or more master LIN nodes to a Controller Area Network (CAN) to obtain the ability of data on one LIN bus to reach other areas in a vehicle not directly wired to a particular LIN. Thus, in this chapter we will move our intra-vehicle communications focus a bit higher, no pun intended, by turning our attention to the operation and utilization of the Controller Area Network. Commencing with an overview of CAN, we will probe deeper into the technology, examining its operations at layers 1 and 2 of the Open Systems Interconnection (OSI) Reference Model, to include the type and format of its frames and the manner by which errors are detected and corrected.

5.1 Overview

The Controller Area Network (CAN) represents a rugged serial bus designed by Bosch in 1986 for use in industrial environments as well as for in-vehicle applications. CAN operates at the first two layers of the Open Systems Interconnection (OSI) Reference Model, which are the physical and data-link layers. Under CAN a broadcast transmission method is employed for placing frames on the bus. One version of CAN supports data rates up to 125 kbps using fault-tolerant transmission at distances up to 40 m, which provides a much higher data transfer capability and an extended transmission range compared to the Local Interconnect Network. In addition, other versions of CAN support data rates up to 1 Mbps, which represents an increase by a factor of 50 over the maximum data rate supported by LIN.

The CAN data-link layer protocol was first standardized by the ISO 11519 standard. This standard was promulgated in 1993. The ISO 11519 standard describes the data-link layer, which is subdivided into Logical Link Control (LLC) and Media Access Control (MAC) sublayers as well as portions of the physical layer. Unlike LIN, which is limited to 16 slaves controlled by a master node, CAN theoretically has the ability to link up to 2032 devices on a single network. However, due to the limitation of hardware transceivers, a CAN from a practical viewpoint can connect up to 110 nodes on a single network.

5.1.1 Evolution

As previously mentioned, CAN was first developed by Bosch in 1986. At that time Mercedes Benz had requested a communications system for its vehicles that would be capable of interconnecting three electronic control units (ECUs). Bosch determined that the UART was not suitable, as it was designed for point-to-point communications. Recognizing the need for a multimaster communications system, Intel Corporation fabricated the first CAN silicon chip, which was followed by the development of controllers from Phillips and other vendors.

The initial Intel silicon chip was the Intel 82526 CAN controller. This chip provided a dual-port random access memory (DPRAM) interface for programming. The 82526 was followed by the Phillips 82C200 CAN controller, which uses a first-in, first-out (FIFO) queue programming model. To distinguish between programming the Intel and the Phillips controllers, some persons referenced the Intel method as full CAN, while the Phillips method is referred to as basic CAN. Today most CAN controllers support both programming methods, which eliminates the need to reference a specific programming term.

5.1.2 CAN Versions

The original Bosch specification evolved over the years and is now subdivided into two parts. Standard CAN uses an 11-bit identifier, while extended CAN uses a 29-bit identifier. Both parts of the specification define the different formats of the message frame. Do not confuse the terms *standard CAN* and *extended CAN* with *full CAN* and *basic CAN*. As a review, the first pair of terms references the use of an 11- or 29-bit identifier in a frame. In comparison, the second pair of terms references the manner by which a CAN controller is programmed.

The subdivision of CAN into standard and enhanced versions resulted in two ISO standards. ISO 11519 has an upper limit of 125 kbps, while ISO 11898 provides for data rates up to 1 Mbps. To specify different maximum data rates, the two ISO specifications differ primarily in their definition of the physical layer.

5.1.3   Types of Controllers

There are three types of CAN controllers: Part A, Part B, and Part B passive. Each of the three controllers supports the standard version of CAN, which uses 11-bit identifiers. However, when used with the 29-bit identifier, a Part A controller will generate errors while a Part B passive controller will be tolerated but ignored. Thus, a Part B controller should be used with extended CAN, as it can send and receive messages in both formats.

5.1.4   Layered Architecture

Similar to LIN, CAN represents a layered architecture. Figure 5.1 illustrates the relationship between the seven-layer International Standards Organization (ISO) Open System Interface (OSI) Reference Model and the CAN protocol stack, with special emphasis on the lower two layers, as the architecture of CAN corresponds to the two lower layers of the ISO Reference Model.

In examining Figure 5.1 note that an embedded CAN controller or independent controller corresponds to both the data-link layer and physical signaling at the physical layer. In comparison, the CAN transceiver, which attaches to the CAN bus, represents the physical medium attachment (PMA) and physical dependent interface (PDI) specifications of the physical layer. Thus, the CAN transceiver specifications include the manner by which pulses are formed as well as connectors and the cable used for the bus. That bus differs from the single-wire LIN bus, as it is a two-wire bus. This two-wire bus has a multimaster capability, which indicates that multiple devices connected to the bus can communicate with one another.

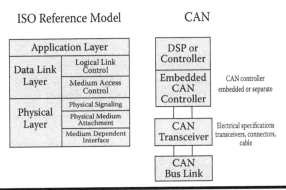

Figure 5.1   Comparing the layered structure of CAN to the ISO reference model.

5.1.5 The CAN Bus

The two-wire CAN bus represents the most popular implementation of CAN. The two-wire CAN bus uses non-return-to-zero (NRZ) signaling with bit stuffing. The term *NRZ* means that the transmission of two successive 1 bits does not result in the signal first being lowered to zero after the first 1 bit.

The left portion of Figure 5.2 illustrates the connection of a CAN controller to a two-wire CAN bus. Note that a CAN transceiver has two connections to the bus. The first connection (CANh) is used to transmit a differential signal, while the second connection (CANl) is used to monitor the CAN bus, which also provides for the receipt of the receiver signal by the CAN controller.

The CAN controller is usually integrated on a digital signal processor (DSP) chip, which in turn is built into an electronic control unit (ECU). Thus, the CAN controller provides the mechanism whereby one ECU can communicate with another to check its status or exchange information. The left portion of Figure 5.2 illustrates the NRZ signaling method used on the CAN bus as well as the relationship between the transmit and receive signaling state and dominant and recessive signals on the bus.

Now that we have an appreciation for the term *NRZ* and the connection of a CAN controller to the two-wire CAN bus, let us turn our attention to the term *bit stuffing*. Bit stuffing prohibits the transmission of a string of six consecutive zero (000000) or six consecutive one (111111) bits by inserting an opposite bit in the data stream to prevent the transmission of six bits set to a 0 or 1. At the receiver an opposite action occurs, with the receiver removing the stuffed bit. Under CAN a bit stuffing violation in which six consecutive bits of the same type are received is considered to represent an error.

CAN uses a modified carrier sense multiple access with collision detection (CSMA/CD) method for nodes to gain access to the bus, in addition to an arbitration process. In a CAN each device listens to the bus to determine if the message flowing on the bus is the same as it is trying to transmit. If it is different, the device

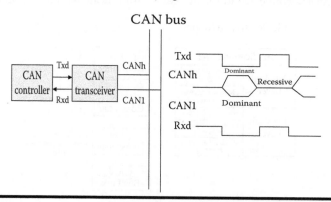

Figure 5.2 The CAN two-wire physical layer and signal on the bus.

will immediately release the bus. This process ensures that one master will always win and results in no messages lost due to a collision.

5.1.5.1   Signaling States

Under CAN there are two different signaling states, referred to as dominant (logical 0) and recessive (logical 1). These signaling states correspond to certain electrical levels, which depend upon the physical layer used. As we will shortly note, there are several different physical layers that can be used by CAN.

At the CAN transceiver the connection to the bus represents a wired-AND function. This means that if just one node is driving the bus to a dominant state, then the entire bus will be in that state regardless of the number of nodes transmitting a recessive state.

5.1.5.2   The Physical Layer

Currently there are several different physical layers defined under the CAN specification. The most common physical layer specification is the one defined in the ISO 11898-2 specification for a two-wire balanced signaling method. This specification is also referred to as high-speed CAN.

Under the ISO 11898-3 specification another two-wire balanced signaling scheme is defined, for lower bus speeds. This is a fault-tolerant specification, which enables signaling to continue even if one bus wire should become cut or shorted to ground or battery. This specification is referred to as low-speed CAN.

A third common physical layer is defined by the Society of Automotive Engineers (SAE) in the J2411 specification. This specification defines the use of a single-wire-plus-ground physical layer, which is used primarily in certain vehicles, such as GM automobiles.

5.1.5.3   Data Transmission

Under CAN, arbitration-free transmission is used to place data on the bus. That is, a CAN message transmitted with the highest priority will satisfy the arbitration while nodes transmitting lower-priority messages will sense the higher priority and back off and wait for access to the bus.

Arbitration-free transmission is supported by the use of dominant (logical 0) and recessive (logical 1) bits. This means that if one node transmits a dominant bit while another node transmits a recessive bit, then the dominant bit wins the arbitration. Table 5.1 indicates the bus state for two nodes transmitting as well as the value of a logical AND between the two.

Table 5.1 Truth Tables for Dominant/Recessive and Logical AND

Bus State with Two Nodes Transmitting			Logical AND
	Dominant	Recessive	01
Dominant	Dominant	Dominant	000
Recessive	Dominant	Recessive	101

Based upon the entries in the truth tables shown in Table 5.1, let us assume one node is transmitting a recessive bit (logical 1) when another node transmits a dominant bit (logical 0). The node transmitting the recessive bit (0) sees the dominant bit (which creates a voltage across the bus while the recessive bit is not asserted on the bus) and determines a collision occurred. Thus, the node transmitting a recessive bit will back off. Then, instead of transmitting, it will wait six bit durations after the end of the dominant message prior to attempting to retransmit.

During the arbitration process each transmitting node will monitor the state of the CAN bus, comparing the received bit with the transmitted bit. If a dominant bit is received when a recessive bit is transmitted the node will stop transmitting.

The actual arbitration process commences during the transmission of the identifier field in the message frame. Each node that commences transmission at the same time places an identifier field with a dominant bit as binary 0, beginning with the high-order bit. As soon as the node ID is a larger number, which indicates a lower priority, the node will transmit a binary 1 (recessive) and observe a binary 0 (dominant), which provides the indication to back off and wait. At the end of the transmission of the identifier field all nodes except the node with the highest-priority message will have backed off, while the node with the highest priority continues its transmission. This action results in the higher-priority message gaining access to the bus, while lower-priority messages will automatically retransmit in the next bus cycle or in a subsequent bus cycle, assuming that there are other higher-priority messages waiting to gain access to the bus.

5.1.5.4   Interoperability Issues

Because different physical layers as a rule are not interoperable, the cost of CAN components, such as transceivers, cannot be amortized over a very large number because different transceivers are used with different physical layers. This is turn drives the cost of CAN upward and results in the use of LIN as a mechanism to provide a lower-cost communications capability for groups of up to 16 slave nodes.

5.1.5.5   Bus Speed

As previously mentioned in this section, the maximum speed of a CAN bus (ISO11898-2) is 1 Mbps, while a low-speed CAN (ISO11898-3) has a data rate up to 125 kbps. In addition, a single-wire CAN has the ability to transmit at a data rate up to approximately 50 kbps in its standard mode of operation, while its high-speed mode allows a data transfer capability of up to approximately 100 kbps. Because the type of transceiver used also governs the obtainable data rate, it is possible that transmissions may have both an upper and lower boundary, as some transceivers cannot transmit below a certain data rate.

5.1.5.6   Cable Length

At a data rate of 1 Mbps a maximum cable length of 40 m, or 130 ft, can be supported. Because the pulse width is inversely proportional to the data rate, slowing the transmission rate widens pulses transmitted on the bus, which in turn enables the transmission distance to be extended. Thus, at a data rate of 500 kbps the maximum cable length is increased to 100 m (330 ft), while at a data rate of 250 kbps the maximum cable length is extended to 500 m (1600 ft). From a practical standpoint, the lower-speed versions of CAN are more suitable for the factory floor where extended cable lengths are required.

5.1.5.7   Bus Termination

Under the ISO 11898 CAN standard the CAN bus must be terminated. The termination of the ends of the bus is accomplished through the use of a 120-ohm resistor at each end of the bus. The use of 120-ohm resistors removes potential signal reflections at the end of the bus as well as ensures that the correct DC level flows on the bus.

Figure 5.3 illustrates the structure of a typical CAN bus. Note that each end has a 120-ohm resistor to remove signal reflections. The actual bus length will vary based upon the data rate of the bus, with higher data rates reducing the length of the bus.

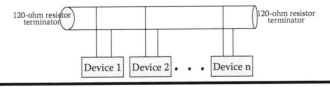

Figure 5.3   A typical CAN bus.

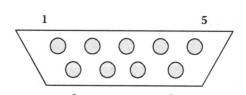

Figure 5.4 Nine-pin DSUB connector and pin assignments: 1 = reserved; 2 CAN_L = CAN_L bus line (dominant low); 3 CAN_GND = CANZ Ground; 4 = reserved; 5 CAN_SHLD = optional CAN shield; 6 GND = optional CAN ground; 7 CAN_H = CAN_H bus line (dominant high); 8 = reserved (error line); 9 CAN_V+ = optional power.

5.1.5.8   Cable and Cable Connectors

Under the ISO 11898 specification a twisted-pair cable that can be either shielded or unshielded is acceptable. Under the SAE J2411 specification a single wire is defined for use.

Although there is presently no standard defined for CAN connectors, the higher layer of the protocol stack defines a few preferred connectors. Three of those connectors are the nine-pin D-sub, five-pin Mini-C, and six-pin Deutsch.

The top portion of Figure 5.4 illustrates the pin positions and assignments for the nine-pin D-sub connector. This illustration represents a male connector viewed from the connector side or a female connector viewed from the sodering side. The power portion of Figure 5.4 contains a table indicating the pins and pin assignments of the connector.

Although the nine-pin D-sub connector is the most popularly used in a CAN, both the five-pin Mini-C and six-pin Deutch connectors are also used. The five-pin Mini-C connector resembles two concentric circles with five pins spaced within the inner concentric circle and is compatible with both standard and extended CAN. The six-pin Deutsch connector is primarily used for mobile hydraulic applications.

5.2   Message Frames

CAN has the ability to support the transmission of four different message types, with each message broadcast on the bus. This means that all nodes literally hear each transmission, requiring hardware to provide local filtering that enables a node to react to messages of interest to the node. The four types of messages that can flow on a CAN bus include:

■ Data frame
■ Remote frame

- Error frame
- Overload frame

5.2.1 Data Frame

The CAN data frame represents the most common type of message transmitted on the CAN bus. The first version of CAN, which is defined by the ISO 11519 specification, uses an 11-bit identifier field that, when combined with a one-bit remote transmission request (RTR) field, is used to determine the priority of messages when two or more nodes are contending for access to the common bus. This version of CAN operates at data rates up to 125 kbps and is referred to as standard CAN.

A second CAN data frame uses a 29-bit identifier formed by adding an 18-bit identifier field to the standard CAN frame as well as incorporating three modifications to the frame, which we will shortly discuss. This type of frame is referred to as an extended frame and can operate at data rates up to 1 Mbps. For an extended CAN data frame the arbitration field, which is employed to determine the priority of messages when two or more nodes contend for access to the bus, consists of a 29-bit identifier field formed by separate 11-bit and 18-bit identifier fields and the RTR bit. Now that we have a basic appreciation for the two types of data frames, let us examine their composition in detail.

5.2.1.1 Standard Data Frame

Figure 5.5 illustrates the fields in the standard CAN data frame. Both the low-speed CAN, defined by the ISO 11519 specification, and CAN 2.0A, defined by the ISO 11898 specification, are compatible with the use of an 11-bit identifier field. The primary difference between the two ISO specifications is the fact that the original standard CAN operates at 125 kbps while CAN 2.0A operates at 1 Mbps.

In examining both Figure 5.5, which illustrates the standard CAN data frame format, and Figure 5.6, which shows the extended data frame format, you will note the absence of an address field. Because CAN messages are broadcast on the bus, there is no need for an address field. Instead, CAN messages can be considered to be content addressed because the contents of a message determine if a node acts upon a message.

Another item worth noting is the fact that the presence of an ACK bit does not indicate that any of the intended nodes have received the message. This bit can be set by any controller that was able to correctly receive the message, by sending an ACK bit at the end of the message. Thus, the ACK bit only informs us that one or more nodes on the bus correctly received the message.

S O F	11-bit identifier	R T R	I D E	r0	0 ... 8 data bytes	C R C	A C K	E O F	I F S

Figure 5.5 Standard CAN data frame format. SOF, a bit that marks the start of a frame and is used to synchronize nodes on a bus after being idle; identifier, an 11-bit identifier that establishes the priority of a message, where a lower value indicates a higher priority; RTR, the remote transmission request bit is set when information is required from another node (although all nodes on the bus receive the request, the identifier determines the node that responds); IDE, the single identifier extension bit is set to define a standard CAN identifier without an extension; R0, a reserved bit for future use; DLC, a 4-bit data length code that indicates the number of bytes of data transmitted; data, 0 to 64 bits (8 bytes); CRC, a 15-bit cyclic redundancy check containing the checksum (number of bits transmitted) of the preceding application data for error detection (in actuality the CRC field consists of a 15-bit CRC and a recessive delimiter bit that indicates the end of the field); ACK, each node that receives an accurate message overwrites this bit position with a dominant bit, which indicates the message was received error-free (if a receiving node detects an error, the message is discarded and the sending node repeats the message; the ACK field is two bits in length, with the first bit used for acknowledgment and the second functioning as a delimiter); EOF, a seven-bit end-of-frame (EOF) field marks the end of a CAN message; IFS, a seven-bit inter-frame separator (IFS) represents the amount of time required by a controller to move a correctly received frame into its message buffer area.

5.2.1.2   Extended Data Frame

The extended CAN data frame, as previously noted in this chapter, uses a 29-bit identifier. To extend the identifier, the extended CAN frame added an 18-bit identifier field, which is separated from the original standard CAN 11-bit identifier field by two fields, a substitute remote request (SRR) field and an identifier extension (IDE) field.

Figure 5.6 illustrates the format of the extended CAN message frame. In comparing Figure 5.4 to Figure 5.6, note that other than the use of an 18-bit identifier field to extend the identifier to 29 bits, the extended CAN data frame only differs from the standard CAN data frame by the addition of three fields:

S O F	11-bit identifier	S R R	I D E	18-bit identifier	r0	r1	0 ... 8 data bytes	C R C	A C K	E O F	I F S

Figure 5.6   The extended CAN message frame.

SRR — Substitute remote request bit, which replaces the RTR bit in the standard message location as a placeholder in the extended frame

IDE — Identifier extension (IDE) bit, which indicates that an 18-bit extension identifier follows

R1 — An additional reserved bit

5.2.1.3   Arbitration

For both the standard and extended CAN frames the arbitration field that is used to determine the priority of a message when two or more nodes contend for use of the bus can be considered to represent a pseudo-field. This field under standard CAN contains an 11-bit identifier and the RTR bit, which is dominant for data frames. Under extended CAN the arbitration field consists of a 29-bit identifier, two recessive bits (SRR and IDE), and the RTR bit.

5.2.1.4   Bit Stuffing

For both standard and extended CAN frames bit stuffing results in the insertion of a bit of opposite polarity after a sequence of five bits of the same polarity occurs. Bit stuffing covers both standard and extended frames from the start-of-frame bit field through the 15-bit cyclic redundancy code field.

5.2.2   Remote Frame

A third type of message that can be transmitted on a CAN bus is the remote frame. The remote frame is similar to the standard and extended CAN data frames. However, there are two key differences between the remote frame and each type of data frame. First, the remote frame has no data field. Second, the remote frame is explicitly marked by the RTR bit being set recessive.

5.2.2.1   Operation

Remote frames can be used to invoke a request–response type of bus traffic. For example, if Node A transmits a remote frame with its arbitration field set to a value of 246, then a node that determines the request frame requires a response would respond with a data frame with its arbitration field similarly set to a value of 246.

Unlike data frames that commonly flow on a CAN bus, remote frames are not commonly used. However, when used, the data length code field must be set to the length of the expected response. If not, arbitration will not work.

5.2.3   Error Frame

A fourth type of frame supported by CAN is the error frame. This frame is transmitted by any node detecting an error. In actuality, the error frame represents a special message that violates the rules of a CAN message. The error frame is transmitted when a node detects an error in a message, causing all other nodes in the network to similarly transmit an error frame. The original node that transmitted the error frame automatically retransmits the message. Through the use of error counters in the CAN controller, which will be reviewed in the next section in this chapter, a node is prevented from continuously transmitting error frames, which in effect would lock up the bus.

The error frame consists of two fields. The first field is the error flags, which is created by the superposition of error flags contributed by different nodes on the bus. There are two types of error flags: active and passive. An active error flag is transmitted by a node that detects an error on the network that is in the "error active" error state. In comparison, a passive error flag is transmitted by a node that detects an active error frame on the network that is in the "error passive" error state.

5.2.4   Overload Frame

The fifth type of frame that can flow on the CAN bus is the overload frame. This frame is transmitted by a node that becomes too busy to process additional data. Thus, the purpose of this frame is to provide for an additional delay between messages.

5.3   Error Handling

In concluding our examination of the operation of the Controller Area Network we will turn our attention to one of the more important aspects of the technology: the manner by which error handling occurs. However, prior to doing so, let us first briefly review how conventional communications technology detects and corrects errors, as this will provide a frame of reference for comparing CAN error handling to common communications error handling.

5.3.1   Communications Error Handling

In a modern communications environment error handling occurs through either the use of parity, when bytes are transmitted independently of one another, or the use of a checksum, when bytes are grouped into a block for transmission.

5.3.2 Parity Checking

Under parity checking an extra bit, referred to as a parity bit, is added to each byte to be transmitted. Parity bit checking can be either odd or even. Under even parity bit checking the parity bit is set to a binary 0 if the number of set bits in the byte to be transmitted is even. If the number of set bits is odd, then the parity bit is set to a binary 1 so that the sum of all set bits is even. Under odd parity bit checking the parity bit is set to a binary 1 if the number of bits set in the byte are an even number and to a binary 0 if the number of bits set in the byte are an odd number.

Under parity checking only a single bit error can be detected. In addition, there is no easy way to correct a byte with a bit error other than visually or by retransmission of an entire document. Due to these problems, most error detection and correction methods evolved through the blocking of bytes and the addition of a checksum to the block that is computed based upon the use of a predefined algorithm.

5.3.3 Block Checking

Under a communications block checking method a fixed number of bytes are used to generate a block. For example, one common communications protocol that employs block checking is the Xmodem protocol. Under the Xmodem protocol 128 bytes are used to form a block. If the last block is only partially filled with data, then the remainder of the block is filled with pad characters (ASCII 127) until the block is filled with 128 characters.

Under block checking an algorithm is applied to each block to generate a checksum that is appended to the block. Thus, the block and its checksum are transmitted. At the receiver the same algorithm is applied to the received data block and a locally generated checksum is computed. The locally generated checksum is then compared to the transmitted checksum. If they match, the data block is assumed to have been received error-free. Then the checksum is removed and the block is sent from the receiver's buffer for processing on the local computer. In addition, the receiver transmits an acknowledement to the sender, which informs the sender that it is okay to send the next data block. If the two checksums do not match, one or more bit errors are assumed to have occurred. Thus, the receiver will transmit a negative acknowledgment to the sender and place the received data block and checksum into the great bit bucket in the sky. The negative acknowledgment serves to inform the sender to resend the data block to include its checksum. Thus, errors are corrected by retransmission.

Although the use of a checksum lowers the probability of an undetected error, such errors can occur when the algorithm used to create the checksum is relatively simple. To reduce the probability of an undetected error, modern communications systems use a polynomial approach to error checking. That is, the bytes in the data block to be transmitted are assumed to represent a long polynomial, which is divided by a fixed polynomial. The resulting quotient is discarded while the remainder becomes the checksum. However, when a polynomial approach is used, the remainder is referred to as a cyclic redundancy check (CRC), which is placed into a CRC field.

Now that we have an appreciation for the manner by which conventional communications systems perform error handling, let us turn our attention to CAN error handling.

5.3.4   CAN Error Handling

Similar to conventional communications systems, an error handling capability is included in the CAN protocol. Under CAN there are five ways that an error can be detected. Two ways operate at the bit level, while the other three operate at the message level. In general, detecting errors in a message appearing on the CAN bus will result in the controller that detected the error transmitting an error flag. The error flag informs other controllers on the bus to discard the current message, in effect eliminating bus traffic. Then, the originating transmitter will retransmit the previously erroneous message. Thus, the error flag can be thought of as similar to a negative acknowledgment, while errors are corrected via retransmission.

5.3.5   Node Removal

One of the more interesting aspects of CAN error handling is the ability of a node to remove itself from the CAN bus under certain conditions. To obtain the ability to determine if a node should leave the bus, each node maintains two error counters. One error counter increments when a transmit error occurs and logically has the name transmit error counter. The second error counter is incremented when a receive error occurs and has the name receive error counter. Because it is logical to expect that a transmitter detecting an error increments its transmit error counter faster than the listening nodes on the bus will increment their receive error counter, because there is a high probability that the transmitter caused a detected error, the transmit error counter value can be used as a threshold for action. That is, once the transmit error counter value reaches a predefined value, the node associated with the counter will first go into an error passive state. When in an error passive state the node will not actively transmit an error flag when an error occurs. Next, the node will then go into a "bus off" state, which means that the node will not participate in any bus traffic.

Table 5.2 CAN Error Detection Methods

Bit monitoring
Bit stuffing
Frame check
Acknowledgment check
Cyclic redundancy check

5.3.6 Error Detection Methods

As mentioned earlier in this section, the CAN protocol defines five methods whereby errors can be detected. Those methods can be categorized as error detection at the bit level and at the message level. In this section we will turn our attention to the manner by which the CAN protocol detects errors. Those methods used by CAN for error detection are summarized in Table 5.2.

5.3.6.1 Bit Monitoring

Bit monitoring is one of two bit-level error detection methods used by the CAN protocol. Under bit monitoring each transmitter on the CAN bus "reads" the transmitted signal level. If the bit level differs from the one transmitted, the node connected to the bus generates a bit error signal.

5.3.6.2 Bit Stuffing

The second error condition to occur at the bit level results from the bit stuffing process. As discussed earlier in this chapter, when five consecutive bits of the same level (0s or 1s) have been transmitted by a node, the node will add a sixth bit of the opposite level to the transmitted bit stream. This process is similar to the zero insertion method used by the High-level Data-Link Control (HDLC) protocol. That method prevents a sequence of six consecutive binary 1 bits from appearing between two flags that define the beginning and ending of a communications frame. When five consecutive 1 bits occur in any part of a frame other than the beginning and ending flag, the sending station inserts an extra 0 bit. When the receiving station detects five 1 bits followed by a 0 bit, it will remove the previously inserted 0 bit, restoring the bit stream to its original value. Thus, under HDLC a false frame is precluded from occurring due to the zero insertion process.

The bit stuffing method utilized by the CAN protocol is used as a mechanism to prevent excessive DC voltage bus buildup. That is, under CAN data is transmitted using non-return-to-zero (NRZ) coding. This coding method means that a sequence of binary 1s results in a high voltage level for the bit duration of all bits, while a sequence of binary 0s would result in no voltage for the bit duration of the sequence of zero bits. Because a long string of binary 1s could result in DC voltage buildup, while a long string of binary 0s could result in a loss of synchronization, bit stuffing under CAN treats sequences of consecutive bits of each polarity the same. This explains why a sixth bit of the opposite polarity is added to the outgoing bit stream when five consecutive bits of 1s or 0s are transmitted by a node. Because bit stuffing changes any long sequence of binary 0s or binary 1s, if more than five consecutive bits of the same polarity occur on a bus, this represents an error condition. The error condition that is signaled is referred to as a stuff error.

5.3.6.3   Frame Check

The frame check is one of three message errors that can occur under the CAN protocol. Because the CAN message has a number of fixed fields that have a range of predefined values, a single value, or a computed value, it becomes possible to check certain CAN message fields. For example, the CRC delimiter, ACK delimiter, and end-of-frame fields have values that can be easily checked. If the CAN controller detects an invalid value in one or more of these fixed fields it will initiate a form error signal.

5.3.6.4   Acknowledgment Check

A second error message that can occur at the message level is referred to as an acknowledgment error. As a review, all nodes that correctly receive a message regardless of the message destination under the CAN protocol are expected to send a dominant level in the acknowledgment slot in the message, while the transmitter places a recessive level in the slot. If the transmitter does not detect a dominant level in the acknowledgment slot, then an acknowledgment error is signaled.

5.3.6.5   Cyclic Redundancy Check

Each CAN message includes a 15-bit cyclic redundancy check (CRC). This CRC is similar to a checksum, but computed by treating the message as a long polynomial, which is then divided by a fixed polynomial. This results in a quotient and remainder, with the quotient discarded and the remainder becoming the 15-bit CRC. Each node performs a similar computation on the transmitted message, using the same fixed polynomial. If a node's computed CRC does not match the transmitted CRC, a CRC error will be signaled.

5.3.7   CAN Controller Operations

Previously, we noted that a CAN controller can increment two counters, one of which corresponds to recognition that a transmitter error occurred, while the second counter is incremented in recognition that a receive error occurred. In actuality, a CAN controller has its mode of operation controlled by the two error counters. In this section we will examine the states that a controller can be in and how the values of the two error counters are used to move the controller from one state to another.

5.3.7.1   Controller States

A CAN controller can be placed into one of three defined error states: error active, error passive, and bus off. The error active state enables messages to be transmitted and received. Thus, this state represents the normal operating mode of a controller. When an error is detected, an error flag is transmitted by the controller.

The second CAN controller error state is the error passive state. A node enters the error passive state when a controller experiences frequent problems when transmitting and receiving messages. Although the controller can transmit and receive messages in an error passive state, it will transmit an error flag when it detects an error when receiving data.

The third CAN controller state is bus off. A controller enters this state if it experiences significant problems when transmitting messages. Once the controller enters the bus off state it cannot transmit or receive messages until it is reset by the host microcontroller or processor.

5.3.7.2   Mode Control

The actual mode of operation of a CAN controller is determined by the contents of the transmit error counter and the receive error counter. The CAN controller will be in an error active mode when both the transmit error counter and receive error counter contents are each less than or equal to 127. If the transmit error counter value is greater than 127 but less than or equal to 255, the CAN controller will be placed into an error passive mode of operation. Only if the contents of the transmit error counter exceed 255 will the CAN controller be placed into a bus off state of operation. When this situation occurs, the CAN controller must be reset by the host microcontroller or processor to be able to resume an operational capability.

5.3.7.3   Counter Updating

Because the contents of the transmit error counter and receive error counters are the mechanism by which a CAN controller resides in a particular state, let us discuss how those counters are updated.

5.3.7.4 Receiver Error Counter

When a receiver detects an error it will normally increment the value of the counter by 1. There are two exceptions to this. One exception occurs when the detected error was a bit error when an error flag or an overload flag was transmitted. If the receiver detects a dominant bit as the first bit after sending an error flag, its receive error counter will be increased by 8.

The second exception occurs when a node detects 14 consecutive dominant bits after sending an active error flag or an overload flag, or after detecting 8 consecutive dominant bits following a passive error flag and after each sequence of 8 additional consecutive dominant bits. When any of these conditions occurs, every receiver on the bus will increment its receiver error counter value by 8.

5.3.7.5 Transmit Error Counter

The operational setting of the transmit error counter is slightly different from that of the receive error counter. Those differences include decrementing the counter value by 1 after the successful reception of a message unless the value was already 0. In addition, the counter value will also be decreased by 1 if it is between 1 and 127 upon the successful reception of a message, to include the successful sending of the acknowledgment bit. In this situation, if the counter value was 0, it will remain at 0, while a counter value greater than 127 will be set to a value between 119 and 127. Another key series of differences between the transmitter error counter and the receive error counter are two situations that cause the transmit error counter to be incremented by 8:

■ When a transmitter sends an error flag
■ When a transmitter detects a bit error while sending either an active error flag or an overload flag

For both of the above situations the transmitter error counter value will be incremented by 8.

5.3.7.6 Error Signaling

Previously we noted the different types of bit and message errors. However, in doing so we deferred until now the details concerning the manner by which different error signals are formed. Thus, let us turn our attention to this important topic.

When a node detects an error it will place an error flag on the bus as a mechanism to prevent other nodes from accepting the erroneous message. The active error flag consists of a sequence of six low or dominant bits. This sequence of six consecutive low bits represents an intentional bit stuffing violation, which will be detected by all other nodes on the bus. Each of the nodes will respond with its own error flag. Once this occurs, the nodes that need to transmit, to include the node that originated the active error flag, will begin their transmissions. This will result in the occurrence of the CAN arbitration process, where the message with the highest priority wins the arbitration process and obtains the ability to transmit its message.

When the CAN controller is in an error passive mode, the error frame will be in the reverse state of an active error flag. That is, it will consist of six passive or high bits. Because the error flag now consists of passive bits, the bus is not affected. Thus, if no other nodes on the bus detect an error, the message will reach its destination without interruption. Note that the error flag is used when a node has recognized a receiving problem that does not require the bus to be affected. Because error handling is automatically performed by the CAN controller, there is no need for the host microcontroller to perform any error handling operations. Thus, the error handling performed by the CAN controller enables the microcontroller to perform other functions.

Chapter 6

Intra-Vehicle Communications

In this chapter we will examine both wired and wireless communications that are primarily used to support intra-vehicle communications. Beginning with a focus on wired technology, we will review the key characteristics of the Local Interconnect Network (LIN) and the Controller Area Network (CAN), with specific focus on the range of vehicle applications they can support. In the second portion of this chapter we will turn our attention to wireless technology, describing and discussing the use of Bluetooth and other wireless technology-based systems that function as delivery mechanisms for information provided by cell phones and satellite transponders.

6.1   Wired Communications

Today there are two primary networking technologies that can be considered to provide a wired intra-vehicle communications capability: LIN and CAN. We examined the operation of LIN in Chapter 4, to include its master–slave relationship and the manner by which data is transmitted on the LIN serial bus. In Chapter 5 we performed a similar examination concerning CAN. Thus, in this section we will first compare the two wired technologies. Once this is accomplished, we will then turn our attention to the applications supported by each networking technology.

Table 6.1 Comparing Key Features of LIN and CAN

Feature	LIN	CAN
Network topology	Bus	Bus
Number of wires	2	1
Maximum data rate	20 kbps	1 Mbps
Communications method	UART based	Controller based
Network access	Master-initiated transmission	Nondestructive
Node support	1 master, up to 15 slaves	64–128; typically limited by physical layer or higher-layer protocol

6.1.1 Network Comparison

Table 6.1 provides a comparison of the major features of LIN and CAN. In the table this author has compared five key technologic features of each networking technology. In doing so, this author avoided listing and comparing other features that, while important for certain applications outside of intra-vehicle communications, such as the maximum transmission distance, which would be important in an industrial operation, are not really beneficial from the view of communications within a vehicle.

In examining the entries in Table 6.1 it is apparent that both LIN and CAN, while both supporting a bus topology, have significant differences. Those differences include the number of wires used, maximum data rate achieved, method of network access supported, and number of nodes that can be in a network. By examining the entries in Table 6.1 it is apparent that CAN is a more robust networking technology that can support four to eight times the maximum number of nodes in a LIN environment at a maximum operating rate. However, to achieve this higher level of performance, CANs are controller based, whereas LINs are universal asynchronous receiver-transmitter (UART) based. This results in the cost of a LIN, to include the chips required to connect nodes to the bus, being significantly less than the cost of the electronics required to attach nodes to the bus in a CAN environment.

6.1.2 Two-Tier Approach

Due to the previously mentioned cost disparity as well as the fact that the lower data rate of LIN is sufficient for most intra-vehicle communications applications, a two-tier approach is evolving with respect to the two technologies. That is, LIN is used to complement CAN, functioning as a "sub-bus" to control several functions

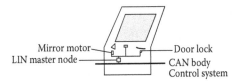

Figure 6.1   Using a LIN sub bus to connect electronics on a car door to other devices via a CAN.

in a localized area, such as a door. By connecting a master LIN node to the CAN, communications between the motor units connected to the LIN and other parts of the vehicle occur.

Figure 6.1 illustrates the use of a LIN sub-bus to provide communications between two motor units and a door lock built into the left door of a vehicle, and other components of the vehicle that can either be directly connected on the CAN or reside on another LIN that is in turn connected to the CAN.

6.1.3   LIN Applications

Vehicle manufacturers have turned to the use of the Local Interconnect Network as a low-cost mechanism to provide a communications capability to electronic units within a localized area. As shown in Figure 6.1, a common approach is to connect electronic units on a door to a LIN, using the master node as a LIN-CAN bus gateway to extend communications to other areas of the vehicle. Although Figure 6.1 indicates only three electronic devices on a door connected as slave nodes on the LIN, the actual number of devices will vary based upon the vehicle and vehicle options. For example, on some vehicles the interior door controls can include separate child window and door locks on the driver's side. When either lock is activated, it overrides the controls on the other doors of the vehicle and can be supported by a sensor connected to a slave LIN node. Thus, connecting the LIN supporting the driver's door electronics to the CAN provides a mechanism to convey the state of the child window and door locks to the controls on the other vehicle doors.

6.1.3.1   Localized Vehicle Area Support

In addition to providing support for the electronics within a vehicle's door, LIN is commonly used to provide a communications capability for other localized vehicle areas, such as the seats in a vehicle, engine and steering controls, the steering wheel and its assortment of controls, and vehicle roof sensors, electronic units, and lights.

6.1.3.2   General Support Areas

Table 6.2 provides a list of typical LIN localized support areas and the functions and electronic controls they may support. Obviously, the latter will vary based upon the vehicle and optional equipment added to the vehicle.

In examining the entries in Table 6.2 it is important to note that the electronics linked by LIN supporting localized areas can considerably vary from one vehicle to another. For example, the electronics integrated into a Mercury Sable wagon do not include the capability to control the radio from the use of buttons on the steering wheel. In comparison, a Mercedes 2007 SLK provides the driver with the ability not only to control the radio from buttons on the steering wheel, but also to control a cell phone via optional wiring integrated into the vehicle's electronics as well as view and reset a variety of trip counters, such as miles per gallon and other

Table 6.2   Typical LIN Localized Support Areas

Doors
Mirror control
Mirror switch
Window lift
Door locks
Engine
Sensors
Small motors
Roof
Moon roof controls
Light sensor
Interior light
Visor lighting
Seat
Occupancy sensor
Seat position motor
Seat heater
Steering column
Wheel tilt/position control motor
Cruise control switches
Wiper control
Turn control

counters. Now that we have an appreciation for LIN applications, let us turn our attention to CAN applications.

6.1.4   CAN Applications

In this section we will turn our attention to some of the more popular and evolving applications supported by the Controller Area Network (CAN). As we begin our examination we will note that unlike LIN, CAN can be used for a wide range of nonautomotive applications.

6.1.4.1   Nonautomotive Support

One of the key differences between LIN and CAN is the fact that the latter is used in numerous applications beyond vehicles. Many industries have adopted the CAN bus standard, and its use ranges from forklifts to connecting subsystems in ships, to providing a backbone network for flight status sensors and navigation systems in aircraft, to its use in factory and building automation, as well as elevators and other types of industrial controls. CAN has even found use in medical systems because laboratory automation is very similar to manufacturing automation. Although each of these application areas is important, we will exclude them from consideration because the focus of this book is upon communications within and between vehicles.

6.1.4.2   Vehicle Operations

In the area of vehicle operations CAN has been used as an in-vehicle network for engine management as well as in applications similar to those previously described as supported by LIN, such as door and roof control, air conditioning, and lighting. However, because LIN represents a lower-cost networking capability, the trend over the past few years is to replace certain CAN-based connections with LIN, connecting the resulting LIN to CAN via a LIN master node. When you consider the fact that saving a few dollars per connection node can be multiplied by 50,000 or more vehicles per vehicle model, the cumulative savings can result in a manufacturer saving several millions of dollars to tens of millions of dollars per vehicle by using both LIN and CAN in a vehicle.

In this section we will primarily focus our attention on CAN applications, to include those that LIN is gradually replacing.

6.1.4.3   Utilization

Developed in Germany, CAN was originally used for power train applications. Currently, CAN is used by a majority of European car manufacturers as the in-vehicle network for engine management, door and roof control, air conditioning, lighting, and even entertainment control. Concerning its use in engine management, several electronic control units are typically connected to a CAN bus operating at 500 kbps to 1 Mbps, which provides a near-real-time communications capability beyond that obtainable on the much slower LIN. In addition, a majority of European-manufactured vehicles use a CAN-based multiplex system, which is used to interconnect various vehicle body electronic control units. Such multiplex networks connect door, roof, lighting, and seat controls and typically operate at a lower data rate than the engine management in-vehicle network (IVN), nominally at 125 kbps. In North America vehicle manufacturers commonly use a single-wire CAN bus-based network in body electronics; however, due to the economics associated in favor of LIN, manufacturers are converting to the lower-cost network as an economy measure.

6.1.4.3.1   Diagnostic Interface

Both European- and North American-based vehicle manufacturers are now implementing a CAN-based diagnostic interface into some vehicles. Most diagnostic interfaces are based on the ISO 15765 standard ("Diagnostics on CAN"), which describes the physical, transport, and application layers as well as the use of what is referred to as Keyword 2000 diagnostic services.

6.1.4.3.2   Power Train Applications

If a vehicle mechanic awoke from a 25-year coma he would more than likely be mystified by the shift from mechanical to electronic subsystems in automobiles. As previously mentioned, CAN was originally developed to support power train applications. When it was introduced, it competed with two other networking technologies: the A-bus developed by the Volkswagon Group and the French Vehicle Area Network (VAN). Today CAN is the survivor, being used in most vehicles manufactured in Europe, North America, and Asia.

In the area of power train applications, emission rules and regulations, coupled with fuel consumption standards and an operator's desire for a quieter ride, can result in a large number of measurements that must be used to dynamically adjust a similar number of engine parameters to provide a desired balance between demands that may be contrary to one another. To achieve this balance requires a closed-loop system where measurements are recorded and used to adjust various engine parameters in near-real-time. Thus, CAN provides the high-speed bus that enables contrary power train applications to be enhanced through the use of CAN communications.

In addition to power train applications, CAN is commonly in use or being considered for use in accident avoidance systems, parking assistance systems, electronic stability control, infotainment systems, and tire pressure monitoring. In some systems CAN provides a full and direct communications support, while other systems provide an interface to CAN that enables automobile manufacturers to integrate independent systems into a vehicle's electronics control display, which is then directly supported by CAN. Because of the importance of each of these applications, we will discuss them in the following paragraphs as separate entities.

6.1.4.3.3   Accident Avoidance System

One solution to avoiding accidents can be obtained by an adaptive cruise control (ACC) system that integrates ultrasound and radar sensors with the cruise control. As a vehicle employing the adaptive cruise control approaches a slower-moving vehicle, it automatically reduces its speed to prevent a collision. Once the vehicle operator takes control of the vehicle and moves it to a lane without another nearby vehicle, the vehicle resumes its set cruise control speed. Thus, the cruise control speed becomes variable and results in the name of the technology.

Currently there are a number of accident avoidance systems either on the market or expected to be released in the near future that differ in operation. For example, Bosch introduced its ACC plus system for low-speed traffic, while BMW's ACC, referred to as Connected Drive, includes a lane departure warning and lane-changing assistance capability. In 2005, Continental launched its Full Range ACC, which was incorporated into the Mercedes S-class vehicles. The Full Range ACC system assists the driver in keeping a set distance from the vehicle in front of his or her automobile. The Continental system was the first to integrate distance sensors at the front of the vehicle with a force feedback gas pedal.

Another ACC system was recently developed by Siemens. This system warns the vehicle operator if the car driving in front of the ACC-equipped vehicle becomes too close. The Siemens system is based on radar sensors that automatically reduce the vehicle speed to maintain a set distance between the two vehicles based upon the speed of the vehicles. Due to its use of radar sensors, the Siemens system is similar to the Bosch system in that both can be considered all-weather systems. Although the Bosch system is already in production, the Siemens system is planned for incorporation into middle-class and compact vehicles during 2008.

6.1.4.3.4   Other Developments

In addition to various automatic cruise control systems that operate in conjunction with radar sensors, some manufacturers have announced plans to incorporate ACC with electronic stability control (ESC), while other equipment manufacturers (OEM) are developing rear-end collision prevention systems. The combination of ACC and ESC would not only increase the performance of ACC, but also save fuel. ACC by itself will keep a vehicle operating smoother in traffic, while ECC enhances deceleration by active braking that does not endanger the vehicle's stability. By combining both via a CAN, radar sensors can be tied to both the ACC and ESC units.

Bosch is working on a system to prevent rear-end collisions by refining its Predictive Brake Assist (PBA) system, which was introduced on the Audi A6 in 2005. Under the PBA system, when ACC radar identifies a critical situation, the brake pads will automatically move closer to the brake disks in the event a potential emergency braking situation becomes a reality. This action gains several fractions of a second if a braking action is required that can be the difference between a collision and a near miss. Because ACC is based upon the use of radar sensors, let us turn our attention to this topic prior to continuing our examination of additional CAN applications.

6.1.4.3.4.1   Millimeter Wave Radar — Millimeter wave radars for years have been used in a range of military and scientific applications for remote sensing and measurement applications. Over the past decade the use of millimeter wave radar has expanded to a variety of commercial applications, to include its use in automobiles as a collision warning sensor.

6.1.4.3.4.2   Types — There are two broad categories of millimeter wave radars, with each category having several modes of operation. The two broad categories are pulsed radar and continuous wave (CW) radar. Table 6.3 provides an indication of the modes of operation associated with each radar category.

6.1.4.3.4.3   Pulse Compression — Both the type of radar and its operational mode determine such characteristics as the size of the radar, its range, and its ability to determine the velocity of a target.

In general, CW radars are simpler than pulse radars and are superior at short-range target detection. In comparison, pulsed radars are preferred for long-range detection. Concerning moving-target discrimination, CW radars can easily determine one target from another, while sophisticated signal processing is required

Table 6.3   Radar Operation Based on Radar Category

Pulsed Radars	CW Radars
Coherent pulsed	Frequency modulated
Doppler	Doppler
Incoherent pulsed	Phase modulated and multifrequency waveform

for pulsed radars to obtain this capability. However, at long ranges pulsed radars become superior in determining target range due to their narrow pulse width. In general, the simplicity of CW radar, as well as its greater ability at short ranges, makes it the preferred type of radar for vehicle use.

6.1.4.3.5   CW Radar Operation

As its name implies, a continuous wave radar transmits a continuous wave signal that is normally frequency modulated. Currently there is interest in using the 60-GHz frequency band for inter-vehicle communications. Due to the high frequency of the previously mentioned band, its use provides a line-of-sight (LOS) condition that enables a vehicle to communicate with other vehicles both in front and behind it. Now that we have an appreciation for the general evolution of accident avoidance systems, to include the use of CW radar, let us return our attention to CAN applications.

6.1.4.3.5.1   Parking Assistance — Commencing with the Lexus 2006 LS400 luxury sedan, we can expect a generation of semiautonomous parking assistance systems to be incorporated into vehicles. Although at first parking assistance can be expected to be incorporated into high-end luxury vehicles, within a few years the technology will be incorporated into mid-range vehicles.

6.1.4.3.6   Operation

Although there are several types of parking assistance systems in commercial use or prototype development, they have certain similarities, which we will discuss. These systems typically use ultrasound sensors mounted on the side of the vehicle to measure parking space length and depth as a vehicle passes the space. The parking assistance system then computes the required steering and informs the vehicle operator visually or acoustically. In some more advanced parking assistance systems the steering will actually be handled by electronically controlled power steering, which is integrated in a closed loop with the ultrasonic sensors. Some parking assistance systems include a miniature TV camera that displays "dead zone" locations on the center console vehicle display, which further assists the vehicle operator. Because the processing of the output of ultrasonic sensors and the display of video result in a data rate that exceeds the capability of a LIN bus, the use of CAN represents an application that will more than likely not be capable of being supplanted by a LIN serving as a parking assistance subunit.

6.1.4.3.7   Body Electronics

After its initial use in power train electronics, CAN was used for connecting power mirrors, power windows, lighting systems, seat control, and steering wheel electronics. Over the years CAN has been incorporated in a range of vehicles, with its use in luxury cars expanded to mid-size and even compact vehicles.

One of the more recent expansions in the use of CAN that deserves mention is the connection of this network to adaptive headlamps. Such headlamps adjust to both different driving and weather conditions, resulting in an enhancement to vehicle safety. Some adaptive headlamp systems provide up to five lighting functions, such as country, motorway/highway, active bend, fog, and cornering. When in the country light mode of operation, the headlamp illuminates the left-hand edge of the road brighter and over a greater range than obtainable with a conventional low beam. In comparison, the motorway lighting function switches on once the vehicle attains a speed of 90 km/hr. The cornering light function is enabled when a vehicle is driven slowly around a bend, while fog lighting occurs when mist limits visibility. To enable the adaptive headlamps to switch their mode of operation, information from other electronic control units is obtained via the CAN in-vehicle network. Such information is used by the headlamp ECU to control the operation of the vehicle's exterior lights.

6.1.4.3.8   Keyless Entry

Although the use of electronic stability control (ESU) and accident avoidance systems has rightfully obtained a large amount of publicity for saving lives, one little known use of CAN is in keyless entry systems. The first keyless entry system was introduced into Mercedes-Benz S-class vehicles in 1999. Jointly developed by Daimler Chrysler and Siemens Automotive, the first keyless entry system, referred to as the Keyless Go control unit, included a CAN interface. Via the CAN interface, a vehicle operator is able to press a start/stop button switch once inside the vehicle. That switch toggles the engine on and off as well as locks the gear selector level in the park position. In addition to the preceding, the Keyless Go control unit will provide the vehicle operator with various system alerts when a problem arises, such as if the key fob is left in the vehicle when the operator exits the vehicle.

Since the introduction of the Keyless Go system in 1999, other vehicle manufacturers have integrated similar systems into their vehicles. Today most vehicle manufacturers offer an integrated keyless locking system with a majority of their vehicles, either as a standard with luxury and mid-sized vehicles or as an option with other vehicles. Although many of these keyless locking systems simply use wireless communications to lock, unlock, or activate emergency vehicle flashing, some systems are more fully integrated with a vehicle's electronics and function similarly to the previously mentioned Keyless Go system.

6.1.4.3.9   Tire Pressure Monitoring

Due to the problems associated with certain tires during the later portion of the 1990s, people noted the importance of tire inflation for road safety and fuel economy. Simply stated, underinflation of tires adversely affects a vehicle's stability as well as the tire's tread life and the vehicle's fuel consumption.

There are several types of tire pressure monitoring systems available to the vehicle operator. At the low end or entry level, a driver can purchase a special type of tire stem valve cap. After ensuring the tire has the correct pressure, the replacement of the existing cap with tire pressure monitoring valve caps will result in the caps appearing green. If the tire pressure decreases by approximately 2 psi, the cap will turn red. This tire pressure monitoring system is obviously very elementary and depends upon repeated inspections of the state of the color on the tire stem valve caps. At the opposite extreme are tire pressure monitoring systems that are integrated with a vehicle's electronics.

6.1.4.3.9.1   Types of Monitoring Systems — There are two types of tire pressure monitoring systems (TPMSs) that are integrated into a vehicle's electronics — direct and indirect. A direct TPMS will deliver real-time tire pressure via sensors located in each tire to the vehicle console's microprocessor. If the tire pressure should fall below a predefined level, a warning will be projected onto the vehicle's display console.

In an indirect TPMS the air pressure is monitored indirectly by monitoring the rotational speed of each wheel. Because an underinflated tire has a smaller diameter than a correctly inflated tire, it will rotate more to cover the same distance as the other tires. The monitoring system notes when one or two wheels are rotating faster than the other wheels and generates an alarm.

The primary advantage of the indirect tire pressure monitoring system is economics. Because most vehicles already have wheel speed sensors for antilock braking, it is very economical to add a TPMS capability to such vehicles. Unfortunately, such indirect TPMSs rely on the user resetting the system when tires are changed or reinflated. Thus, if the driver forgets to reset the indirect system, he or she may receive a potentially misleading alert message.

6.1.4.3.9.2   Direct TPMS Example — One example of a direct TPMS integrated into a vehicle's electronics uses battery-powered wheel electronics that periodically measure both the air pressure and temperature inside each tire. Both the tire pressure and temperature inside the tire are transmitted along with battery life and the identification of the wheel electronics via a radio frequency (RF) signal to an antenna mounted in the wheel arch. From the antenna information is relayed by cable to an electronic control unit that examines the data and uses predefined thresholds to determine if the driver should be alerted.

Although the previously described tire pressure monitoring system uses a combination of wireless and wired communications, data flows to an ECU via a wired cable. Direct tire pressure monitoring systems monitor each tire separately. Such systems allow the car manufacturer and driver to input the normal tire pressure to enable the system to note differences. Although factory set data is normally sufficient, if a driver replaces the original equipment manufacturer (OEM) tires at a later date, the ability for the vehicle operator to enter a new series of tire pressure parameters is important. However, to prevent an inadvertent error, such as the transposition of 36 to 63, most systems check the entered values to ensure they are plausible.

6.1.4.3.9.3   Evolution — The first passenger vehicle to have a TPMS was the Porsche 959 in 1986. Later, due to the importance of vehicle safety, tire pressure monitoring systems began to be incorporated into luxury European automobiles, such as the Mercedes S-class and the BMW 7-series. In 1999 Peugeot adopted tire pressure monitoring as a standard feature in its Peugeot 607. In 2000 Renault added TPMS as a standard feature to its Laguna II, the first high-volume passenger vehicle to be sold with a TPMS as a standard feature. During the late 1990s the Firestone tire recall resulted in the Clinton administration publishing the TREAD Act, which mandates the use of tire pressure monitoring technology to alert drivers of severe underinflation of their tires. Based upon the TREAD Act, the National Highway Transportation Safety Administration (NHTSA) has tried three times to issue rules that require a tire pressure monitoring system to eventually be installed on all vehicles sold in the United States. The latest rule, issued in April 2005, would require automobile manufacturers to install pressure monitoring systems in all new passenger cars and trucks by the 2008 model year, beginning with a phase-in with 2006 model year vehicles. Because this rule did not satisfy the requirements set by the U.S. Congress, to include allowing occupants to ride on dangerously underinflated tires, nor did the rule require tire pressure monitoring systems to operate with replacement tires, several tire manufacturers and Public Citizen, a nonprofit public interest organization, filed suit in June 2005 in the U.S. Court of Appeals for the District of Columbia, arguing that the rule is inadequate and should be overturned. Although this request is being evaluated, automobile manufacturers continue to add a TPMS capability to both near-luxury and luxury vehicles. Today such vehicles manufacturers as GM's Cadillac and Pontiac divisions, BMW, Audi, and Mercedes offer a TPMS capability on certain vehicle models and are expected to increase the availability of this feature to other models in their product lineup.

6.1.4.3.9.4   Operational Example — The most common implementation of a tire pressure monitoring system occurs through the combined use of wireless and wired technology. A typical TPMS uses four or five (to include the spare tire) transmitter sensors, one or four RF receivers, and a cable routed from the RF receivers to an electronic control unit. The use of RF transmitter sensors allows manufacturers to avoid expensive and complex rotating contact wiring for each wheel. The transmitter sensor periodically monitors tire pressure and temperature, transmitting the data, to include battery power, in the sensor to the antennas, where the data is relayed via a shielded cable to the ECU. Most TPMS sensors use a lithium battery, a silicon-based pressure sensor, and an RF oscillator, with the silicon chip representing the key component of the sensor as it manages the different components of the sensor.

The use of battery-powered sensors results in the battery life being a critical component of the TPMS. Although the intended life is 10 years, extreme weather conditions can significantly shorten battery life. Due to this, several vendors are developing batteryless sensors, and Michelin announced that it would use such sensors on its Etire II commercial tires.

From the RF antenna a combined wireless–wired TPMS uses a shielded cable to connect the antennas to a TPMS ECU. Some TPMSs use four individual antennas, while other systems use a single antenna. The ECU typically can be obtained with an optional CAN interface. The purpose of the CAN interface is to enable the transmission of preprocessed tire pressure measurements to other ECUs in the vehicle. Thus, information from the tire pressure monitoring unit can flow to the diagnostic ECU in a vehicle, allowing a technician, after making a simple connection to the unit, to obtain a variety of information about the vehicle, to include its tire pressure measurements, without the need to directly connect his or her diagnostic equipment to the tire pressure monitoring system ECU. This obviously reduces the time needed by a technician to check the status of a vehicle.

The TPMS ECU commonly consists of a microprocessor, memory, CAN bus interface, and diagnostic interface. The ECU evaluates data received from each tire to determine if the data falls within a predefined range of values. If the data falls outside the predefined range of values, the ECU will transmit an alert message that will be displayed on the console. Otherwise, the ECU functions transparently with respect to the vehicle operator. If the ECU is connected to a CAN, then diagnostic information can be obtained from a central vehicle diagnostic module instead of from a technician accessing the TPMS ECU.

6.1.4.4   Infotainment

As discussed earlier in this book, the term *infotainment* is used to reference the electronics and displays in a vehicle that include integrated conventional and satellite radio receivers, CD and DVD players, and a navigation system. Several manufacturers of one or more of the previously mentioned components or the entire system include a CAN interface within the infotainment ECU. One example is the Bosch CAN-based system.

Although the standardization of an audio/video bus for vehicle applications may be several years away, in the interim many vehicle manufacturers have selected the Media-Oriented Systems Transport (MOST) multimedia network as a viable candidate for standardization. Because some vendors provide vehicle manufacturers using MOST with an audio gateway to CAN, this gateway makes it possible for information from the CAN-based in-vehicle network to be passed to such infotainment devices as the radio, CD changer, and DVD navigation display. Now that we have an appreciation for intra-vehicle wired communications, let us turn our attention to wireless communications.

6.2   Wireless Communications

Within the past decade the use of wireless technology within a vehicle has literally exploded. Today there are millions of vehicles that use a range of wireless technologies to enable such applications as hands-free cell phone operation, remote door unlocking, updated navigation, and traffic reporting. Although some wireless communications technologies extend beyond a vehicle, such as satellite communications, because communications flow directly into the vehicle we will consider them to represent a wireless intra-vehicle communications method. Thus, in this section we will turn our attention to Bluetooth and such satellite services as satellite radio and satellite-based vehicle services.

6.2.1   Bluetooth

Bluetooth represents an industry standard for the creation and operation of wireless Personal Area Networks (PANs). A PAN is similar to a low-speed, short-range wireless LAN that uses frequency hopping; however, there are some significant differences between the two. In this section we will examine the key parameters of the standard as well as note the similarities and differences between Bluetooth and wireless LANs and some of the intra-vehicle applications supported by Bluetooth technology.

6.2.1.1   Evolution

The Bluetooth specification was originally developed by Ericsson in Lund, Sweden, during 1994. A Special Internet Group (SIG) took over the development of the Bluetooth specification in 1998. Since then, over 6000 companies have joined the Bluetooth SIG and the IEEE has standardized the technology as the IEEE 802.15.1 standard.

Early versions of Bluetooth, such as 1.0 and 1.0B, had many interoperational problems that precluded its adoption by a significant base of products. Later versions of Bluetooth, such as Bluetooth 1.1 and 1.2, fixed many previously reported problems, while version 1.2 included several enhancements that resulted in the wide adoption of the standard, being integrated into cell phones, computers, and vehicle radio systems. Version 1.2 included an adaptive frequency-hopping spread-spectrum (FHSS) capability. This capability significantly improved the resistance of Bluetooth devices to radio frequency interference by avoiding the use of crowded frequencies in the frequency-hopping sequence. This enhancement was of particular importance as wireless LANs use the same radio frequencies as Bluetooth, but with higher power that could "mask" an attempted Bluetooth connection.

In 2004, Bluetooth 2.0 was released, resulting in an extension of transmission range to 100 m at data rates up to three times those obtainable under Bluetooth 1.2. While backward compatible with version 1.2, this new specification also improved the bit error rate transmission performance as well as reduced the Bluetooth device's power consumption.

6.2.1.2   Classes

Bluetooth represents a low-power-consumption communications protocol that enables devices to exchange data with one another. Currently there are three defined Bluetooth classes that designate the maximum permitted power in mW and dBm, and range in meters. Table 6.4 summarizes the three classes. Note that Class 1 requires support of Bluetooth 2.0.

In examining the entries in Table 6.4 you will note that the most powerful Bluetooth class is Class 1, as it provides a maximum power level that can extend the range of transmission to approximately 100 m. For most in-vehicle operations, Class 2 Bluetooth should be more than sufficient, as it enables a transmission range of up to approximately 10 m.

6.2.1.3   Operation

A Bluetooth device functioning as a master can communicate with up to seven devices functioning as slaves. The resulting network formed by one master and up to seven slaves is referred to as a piconet. Technically, a piconet represents an ad hoc network of devices using Bluetooth technology that come within range of one another and automatically form a network. The restriction concerning the number of active nodes in a piconet results from the use of a three-bit Media Access Control (MAC) address to identify active devices, resulting in a maximum of eight devices

Table 6.4   Bluetooth Classes

Class	Maximum Permitted Power (mW)	Maximum Permitted Power (dBm)	Range (m)
Class 1	100	20	100
Class 2	2.5	4	10
Class 3	1	0	1

(one master and seven slaves) in a network. However, up to 255 future slave devices can be inactive or parked, with the master device able to bring inactive devices into an active status whenever necessary as long as the total number of active networked devices does not exceed eight.

Bluetooth supports both unicast and broadcast transmission. Under unicast transmission the master transmits data to slaves in a round-robin manner or a slave can directly communicate with the master node. Although broadcast transmission is supported, from the practical standpoint devices supporting Bluetooth have little need to receive data sent to other devices, resulting in this capability being limited in use.

6.2.1.4   Spectrum Utilization

As previously discussed, Bluetooth uses the same frequency band as wireless LANs. Specifically, Bluetooth operates in the 2.4-GHz band, which is referred to as the industrial, scientific, and medical (ISM) unlicensed frequency band. Bluetooth devices communicate with one another by frequency hopping. A total of 79 frequencies, 2.402 to 2.480 GHz, displaced by 1 MHz are used in the United States. In other countries the frequency band range may be reduced, resulting in a lesser number of frequency hops available for use.

The Bluetooth protocol can change channels up to 1600 times per second. Bluetooth 1.1 and 1.2 have a maximum data transfer capability of 723.1 kbps, while version 2.0, which incorporates an enhanced data rate capability, can achieve a data rate of 2.1 Mbps. Because voice can be digitized at 8 kbps or even at a lower data rate, the use of Bluetooth for connecting a cell phone into a vehicle's audio system does not even begin to tax its data transfer capability.

6.2.1.5   Modulation

Bluetooth uses Gaussian frequency shift keying (GFSK) modulation. Under GFSK a binary 1 is represented by a positive frequency deviation, while a binary 0 is represented by a negative frequency derivation.

6.2.1.6   Frequency Hopping

Similar to one of the methods of transmission employed by wireless LANs, Bluetooth supports a pseudo-random frequency-hopping sequence. The hopping sequence for each piconet is determined by the Bluetooth device address of the master and provides a unique sequence of frequencies that the master and slaves use to communicate, explaining how two piconets can exist side by side without interfering with one another.

Table 6.5   Bluetooth Logical Channels

Channel	Use
LC	Control channel
LM	Link manager
UA	Asynchronous user information
UI	Isosynchronous user information
US	Synchronous user information

The phase in the hopping sequence is based upon the master Bluetooth node's clock. The channel is then subdivided into time slots, with each slot (625 ms in duration) corresponding to a radio frequency hop frequency. Thus, a series of consecutive hops will correspond to different RF hop frequencies.

6.2.1.7   Logical Channels

Bluetooth supports five logical channels that can be used to transfer different types of information. Table 6.5 indicates the logical channels supported by Bluetooth and their use.

6.2.1.8   Device Addressing

There are four types of addresses that can be assigned to a Bluetooth device. One address that was previously mentioned is a three-bit number referred to as the MAC address. This address is valid only when a slave is active on a channel. The other Bluetooth addresses include a device address, parked member address, and access request address.

The Bluetooth device address is a unique 48-bit address. The parked member address is an eight-bit address used to identify up to 255 parked slaves. This address is only valid as long as the slave is parked. The access request address is used by a parked slave to determine the slave-to-master half slot in the access window it can use to transmit access request messages. Similar to the parked member address, the access request address is only valid when the slave is parked.

6.2.1.9   Packets

Bluetooth devices communicate with one another through the transmission of packets within a piconet. Figure 6.2 illustrates the Bluetooth packet format, which consists of an access code, header, and payload. The access code is used for timing synchronization, while the header defines the type of information in the payload or

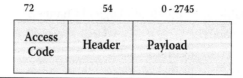

Figure 6.2 Bluetooth packet format.

a predefined function, such as packet acknowledgment, flow control, slave address, and error check. The payload can contain either digitized voice, data, or both.

6.2.1.10 Operational Modes

A Bluetooth device is in one of two major states: standby and connection. In addition, a Bluetooth device can be in one of seven substates at one time, where the substates are used to add slaves or to make connections in the piconet.

The standby state represents the default low-power state in a Bluetooth device. In this state there is no interaction between devices. To obtain connectivity between devices, one device (master) will transmit an inquiry packet that the destination (slave) will respond to with an inquiry response. Once this is accomplished, the source will page the destination. In response, the destination will transmit a slave response. The master will then transmit a frequency hop sequence to the slave, allowing the two devices to exchange information as they hop through a pseudo-random sequence of frequencies.

As devices exchange information, a series of functions are performed that are transparent to the device operations. Those functions include three types of forward error correction schemes to reduce transmission errors, flow control to prevent the loss of data as queues fill, and synchronization of devices to the master's clock.

6.2.1.11 Service Discovery Protocol

In concluding our brief overview of the operation of Bluetooth we will turn our attention to the mechanism whereby applications can discover which services are available and the characteristics of those services. That mechanism is provided by the Service Discovery Protocol (SDP).

The Bluetooth SDP is based upon a request–response model where each request elicits one response. Because there can only be one unacknowledged request at any time, the SDP provides a simple form of flow control. That is, to delay responses only requires the delay of requests.

The Bluetooth SDP can be considered to represent a client–server interaction model. Under this model client applications communicate with the SDP client while server applications communicate with the SDP server. Then the interaction between client and server is provided via the exchange of SDP requests and

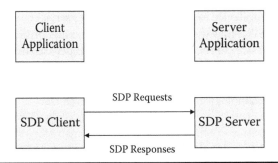

Figure 6.3   The Bluetooth client–server model.

responses. Figure 6.3 illustrates the Bluetooth client–server model in the form of a block diagram.

Although the exchange of SDP requests and responses shown in Figure 6.3 appears simplistic, in certain cases an SDP request can require a response that cannot fit within a single packet. When this situation occurs, the server will generate a partial response as well as a continuation state parameter. That parameter will then be normally used by the client in a subsequent request to retrieve the next portion of the response. Note that because client requests are necessary to retrieve each portion of a multipart response, this does not impact the ability of SDP requests to provide flow control.

6.2.1.11.1   Services

The reason behind the SDP is to enable Bluetooth devices to discover the services available and offered by other such devices. To support this capability, each Bluetooth device has one or more service attributes, each of which describes a characteristic of the service offered. Each service attribute consists of two components: an attribute ID and an attribute value. A complete list of service attributes forms a service record that can define a device's hardware, software, or a combination of both characteristics.

The attribute ID represents a 16-bit integer that identifies the service attribute within a service record. In comparison, the attribute value is a variable-length field whose value is based upon the attribute ID and the service class of the service record in which the attribute is contained.

Under the Bluetooth SDP, a device can search for a specific service or browse to determine what services are offered. If you have a cell phone with a built-in Bluetooth communications capability and a Bluetooth-compatible laptop or desktop, once you are within communications distance, and assuming both devices have Bluetooth and Bluetooth discovery enabled, one device can be used to provide a list of services on the other device by displaying an attribute called the Browse Group List. This attribute contains a list of Universal Unique Identifiers (UUIs) that enables a client device to browse an SDP server's services, which explains how your Bluetooth device may display such services as faxing, printing, and uploading pictures as your device comes within range of other Bluetooth devices that offer different types of services. Concerning such services, because they actually represent applications, we will conclude our overview of Bluetooth by discussing some of the applications the technology could support within a vehicle.

6.2.1.12 Vehicle Applications

There are several existing and potential Bluetooth applications that can be used within a vehicle. The most common type of application involves the extension of the use of a cell phone to a hands-free environment.

6.2.1.12.1 Hands-Free Cell Phone Use

The most common use of Bluetooth within a vehicle is for the driver or passenger to use a hands-free headset that either hangs on one ear or is fitted over both ears and is used to wirelessly control many cell phone operations. For example, by pressing a button on the headset, the driver or passenger can answer an incoming call without having to touch the cell phone. Similarly, pressing the button a second time will disconnect the call. A microphone that extends below the ear allows the driver or passenger to communicate with the caller.

A second type of hands-free vehicle communication involving the use of a cell phone integrates the vehicle's radio system with the phone. Through the use of a Bluetooth adapter typically mounted in the interior of the vehicle's central console, a Bluetooth-compliant cell phone is integrated into the vehicle's electronics. This integration allows the driver or operator to use buttons on the radio to call another telephone, with the radio speakers used to provide audio reception of the distant party. When an incoming call is received, any sound generated by the prior selection of a radio or CD player is temporarily muted, allowing the speakers to be used for the audio generated by the caller. Taking this capability to a higher level, some vehicle manufacturers that support high-end cell phones enable the driver or passenger to use his or her voice to initiate and disconnect calls, as well as to utilize additional such functions as adding and editing the phone book entries in the cell phone.

6.2.1.12.2 Evolving Applications

Because Bluetooth represents a low-cost, limited-transmission-distance technology, it is well suited to enable mobile devices to be integrated into a vehicle's prewired infotainment system, to include the vehicle's console. This means that with appropriate software it will be possible for mobile navigation systems to be moved from one vehicle to another and use the console display to provide a map of the route to a specific destination, to allow the key pad and phone list of a cell phone to operate a mobile navigation system, to enable debit cards to be used to pay tolls via their insertion into a reader integrated into a miniature transponder located within a vehicle, and many other applications. Thus, the use of Bluetooth technology within a vehicle will only be limited by one's imagination in connecting different mobile devices within a vehicle.

6.2.2 Satellite Services

In concluding this section on wireless applications with a vehicle, we would be remiss if we did not mention satellite services. Thus, in this section we will briefly mention several existing and evolving satellite-based applications that are marketed for use within a vehicle, while deferring a more in depth discussion to the next chapter.

6.2.2.1 Satellite Radio

Perhaps the most popular satellite-based service is satellite radio, which provides subscribers with a large number of advertiser-free stations to listen to. Although most satellite-capable radios are built into new vehicles and customers receive a free trial subscription, you can also purchase mobile satellite radio receivers that can be easily moved from a vehicle into a home or office. The latter makes more sense for a subscription-based service, because only a small portion of the day is spent in most vehicles.

Although satellite-based radio has received a majority of the publicity associated with the use of satellites to transmit information to moving vehicles, it is not the only service used to transmit data into vehicles. Two additional satellite-based services that warrant attention are vehicle care- and traffic status-based services.

6.2.2.2 Vehicle Care

The term *vehicle care* covers a range of services, to include vehicle accident notification, emergency door unlocking, and even vehicle tracking, the last designed to lower insurance rates, as it reduces the probability of a stolen vehicle not being recovered in a timely manner. Perhaps the most recognized vehicle care subscription service is General Motor's OnStar system. Since its introduction by General Motors in 1996 as an automotive safety tool, over 5 million persons have subscribed to this service in addition to the subscribers that receive a one-year free subscription when they purchase a new OnStar-capable vehicle.

6.2.2.2.1   Overview

OnStar and competitor services represent a telemetrics service. Here the word *tele-metrics* represents a combination of telecommunications and informatics that provides information to mobile devices, such as PDAs, cell phones, laptop computers, and vehicles. When discussing the operation of a vehicle care system such as OnStar, the term *telemetrics* is used to describe a vehicle's electronics that use a global positioning system (GPS) satellite system for positioning with a cellular technology for communications between the vehicle and the subscription headquarters.

A vehicle equipped with OnStar has a panel with a series of buttons that is located in the rearview mirror, on the dashboard, or on an overhead console. A blue button, when pressed, allows the driver to directly communicate with a live or virtual advisor. A red button with a cross, when pressed, denotes an emergency, while a white button with a handset is used to make a phone call.

6.2.2.2.2   Evolution

OnStar was initially developed as a location-based service by a team at Extreme Blue, an IBM internship program for graduate students. General Motors introduced OnStar service in 1996 as an option on some of its Cadillac models.

Until 2000, OnStar consoles included a handset, with a hands-free console becoming standard in 2001. The cellular system used in OnStar was originally designed for the Analog Mobile Phone System (AMPS). Today most OnStar systems are now dual mode; however, because AMPS is expected to be both obsolete and turned off by 2008, future GM vehicles equipped with OnStar will use all digital cellular systems beginning with the demise of AMPS.

Currently OnStar is available on over 50 models of GM vehicles. Other vehicle manufacturers, to include Audi, Acura, Isuzu, and Volkswagen, also offer OnStar technology. In February 2006, OnStar announced plans to incorporate a real-time navigation system called Turn by Turn in select Buick and Cadillac models. By 2010, Turn by Turn is expected to be a standard service on all vehicles equipped with OnStar.

6.2.2.2.3   Operation

As a telemetrics system OnStar combines electronics with a GPS antenna and a code division multiple access (CDMA) cellular antenna to enable safety information and driver queries to be communicated from a vehicle to an OnStar center. For example, if an accident occurs and air bags are deployed, the OnStar system will note the location of the vehicle via its GPS receiver and initiate a call to an OnStar center, which will convey vehicle data and the user's GPS location, enabling OnStar personnel to both attempt to communicate with the vehicle operator and notify applicable law enforcement or medical authorities. Similarly, when a driver presses the red OnStar emergency or blue OnStar button, current vehicle data, to include the vehicle's GPS location, is sent to OnStar. Presently, there are four OnStar centers, located in Troy, MI, Charlotte, NC, Oshawa, ON, and Makati, Phillipines. Each center is open 24/7 throughout the year.

OnStar uses a 3-W cellular system, which is five times more powerful than the typical portable cell phone, explaining why the clarity of a voice call made via OnStar is a significant improvement over a call made using a regular cell phone. In addition, as long as battery power is available to an OnStar-equipped vehicle, a crook who snaps off the antenna will not disable the built-in cellular system. Although the range of the cellular system will be diminished, once the vehicle is reported stolen, there is a high degree of probability that OnStar can track its movements as well as notify appropriate officials concerning its location.

6.2.2.3   Traffic Status

In concluding our discussion of wireless communications within a vehicle, we will briefly discuss an emerging service that this author will refer to as traffic status.

In some metropolitan areas in the United States traffic tie-ups, road congestion, and accident information may be conveyed to vehicle operators through the use of traffic information displays or signs informing the vehicle operator to turn his or her radio to a certain frequency. Although this manner of information dissemination typically does a good job of informing a vehicle operator about a problem, it provides limited information about alternative courses of action. For example, a driver commuting to a downtown office location more likely requires an alternate route in the event an accident occurs on the main downtown expressway than another driver whose destination is a suburban airport. Recognizing this problem, some vehicle manufacturers have added a concierge service to their telemetrics offering, while several independent companies have begun operations allowing subscribers to call via their cell phone for alternate routing to a specific destination. Regardless of the name of the service, the service recognizes that different vehicle operators typically tied up in a traffic bottleneck have different destinations and an alternate route for one driver may not be suitable for the next driver.

As concierge services and traffic status reporting evolve, this author believes that the next logical step would be to integrate the vehicle navigation system into the cell phone based service. Then, the service provider could download an applicable alternate route into the vehicle operator's navigation system, with the navigation system providing the driver with explicit instructions concerning using the alternate route selected by the subscription service.

Chapter 7

Inter-Vehicle Communications

Although the term *inter* is normally used to reference or describe the flow of data between devices, this author has taken the liberty to expand upon its meaning. In this chapter we will use the term *inter-vehicle communications* to describe the flow of communications between vehicles as well as from a vehicle to sensors or transponders located on a highway, road, or street that is used by a vehicle.

Over the past decade a considerable amount of research has occurred in the area of inter-vehicle communications oriented toward providing safety features and expanded capabilities to vehicles. Such research is oriented toward the use of Mobile Ad Hoc Networking for vehicle safety, enabling groups of vehicles traveling in clusters on a road or highway to dynamically form a network through which different types of information can be relayed from vehicle to vehicle. A second area of research that also warrants attention is the development of the intelligent road, where a combination of signs, displays, sensors, and transponders can provide information to vehicles and vehicle operators. Because an ad hoc network enables information to be relayed from one vehicle to another that forms the network, a combination of ad hoc networking and intelligent roads can be used to extend the range of safety and other information provided by roadway devices.

Based upon the fact that inter-vehicle communications and intelligent roads represent evolving technologies, the purpose of this chapter is to acquaint the reader with each technology. In the first section, we will turn our attention to ad hoc networking, describing the technology and its potential use in inter-vehicle communications. In doing so, we will discuss and describe some of the activities and applications that can be expected to be promoted by inter-vehicle communications that will enhance road safety. In addition, we will also discuss the different types of communications technology that may be used for ad hoc networking as well as the frequency bands they use.

In the second section, we will turn our attention to the intelligent road. In doing so, we will examine how intelligence can be provided to vehicles that use highways, roads, and streets. In addition, because ad hoc networking can be used to extend the transmission range of information provided by intelligent roads, we will also discuss the relationship between the two technologies.

7.1   Ad Hoc Networking

In this section we will focus our attention upon obtaining an understanding of what the term *ad hoc networking* means when applied to vehicles. To do so, we will examine how such networks operate, the rationale for their use, some of the applications they can support, the probable frequency bands they will use, and other technology aspects behind this method of networking.

7.1.1   Overview

An ad hoc network represents a network that has either no infrastructure or a minimal infrastructure. This type of network consists of nodes that come together to form a network; hence, we can say that an ad hoc network is self-organizing. Each node can function as a network router, data source, or data destination. Thus, as two or more nodes come together to form an ad hoc network, they become capable of relaying information as well as communicating with one another.

Based upon the preceding, a vehicle ad hoc network represents the use of vehicles as network nodes. Here the nodes move at will relative to one another but within the constraints of the roadway.

7.1.2   Formation

Figure 7.1 illustrates an example of the formation of a Vehicle Ad Hoc Network. The left portion of the referenced figure shows a highway with three vehicles traveling in the passing lane, while two vehicles are shown traveling in the nonpassing lane, resulting in five vehicles within close proximity to one another that form a Vehicle Ad Hoc Network. In this example the first vehicles in the passing lane and slow lane are assumed to be within communications range of one another. In addition, the first two vehicles in the passing lane are also assumed to be within communications range of one another, while the second vehicle in the passing lane is within communications range of the second vehicle in the slow lane. Finally, the

Figure 7.1   Forming a vehicle ad hoc network.

second vehicle in the slow lane is assumed to be within communications range of the third vehicle in the passing lane.

In the right portion of Figure 7.1, the vehicles are treated as nodes, illustrating the mesh structure that can be used to relay information from one member of the ad hoc network to another member. Note that for simplicity, the vehicle nodes were labeled numerically. In actuality, each vehicle will more than likely use wireless local area network (LAN) technology as a mechanism to form a Vehicle Ad Hoc Network. Because the IEEE 802.11 LANs use 48-bit Media Access Control (MAC) addresses to identify nodes, those addresses would be used to form the ad hoc network. However, for simplicity of illustration, we will use digits in this chapter to represent ad hoc networking nodes.

Turning our attention to the right side of Figure 7.1, note that information between nodes can flow either directly from one node to another or indirectly via an intermediary node. Concerning the latter, data from node 1 can reach node 5 either via nodes 2, 3, 4, and 5, or via nodes 4 and 5. Thus, as the number of nodes in an ad hoc network increases, there will usually be an increase in the number of paths between nodes.

As vehicles move on the highway, the connections between vehicles will dynamically change. For example, if the cluster of vehicles shown in Figure 7.1 results in the first vehicle in the passing lane zooming ahead of the other vehicles while the second vehicle moves within range of the first vehicle in the slow lane, the resulting ad hoc network might be reduced to four nodes, as shown in Figure 7.2. When this situation occurs, vehicle mobility will result in messages having to be routed on different paths to compensate for broken links. In this particular situation, the four vehicles shown in the left portion of Figure 7.2 are situated such that there is now no path between nodes 1, 3, 4, and 5 and node 2, as node 2 is now out of range of the ad hoc network.

7.1.3   Rationale for Use

There are several reasons for considering the use of an ad hoc vehicle network to provide an inter-vehicle communications capability. Perhaps the key reason results

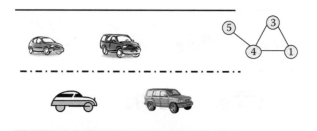

Figure 7.2   Vehicle mobility can result in broken links, which dynamically rearranges the paths between nodes.

from its name (ad hoc), which indicates the lack of a requirement for having an infrastructure available for use. Thus, roadside sensors and transponders do not have to be placed at fixed intervals, but can be clustered at relevant locations for the information they convey to be broadcast for pickup by Vehicle Ad Hoc Networks that come within range of roadway devices.

Another reason for the use of ad hoc networking is the fact that only compatible vehicles that support the communications technology will form the network. This means that the technology can be introduced gradually over a period of years or even decades, and does not depend upon all vehicles on the road having an ad hoc networking capability. In fact, there will probably never be a period in which all vehicles support a new technology, regardless of the benefits of the technology. This is due to the large number of collector vehicles that owners periodically drive, such as Ford Model T's, Studebaker and Avanti cars, as well as more modern Corvettes and Ferraris. Because the value of such vehicles is based upon them being restored to their original condition, it is doubtful that owners would modify them even if low-cost ad hoc networking adapters were available for purchase.

7.1.4   Applications

The range of applications that can be supported by the use of ad hoc networking is basically limited by one's imagination. For example, at the beginning of each workday, there are tens of millions of commuters who use their vehicles to commute to work, while at the end of the workday an opposite commute occurs as vehicles typically exit urban areas for the suburbs. This persistent traffic flow into and out of urban areas commonly results in periods of congestion that is both costly and frustrating to each vehicle operator. The congestion is costly as drivers sit in their vehicle consuming fuel and wasting time, while frustration occurs from the snake-like progression of traffic through commonly unsynchronized lights and the need to actively monitor vehicle traffic around the periphery of one's vehicle as the operator enters or exits a highway, changes lanes, or deals with controlling the vehicle's gas pedal and brake as he or she navigates the daily commute. Based

Table 7.1   Potential Ad Hoc Vehicle Network Applications

Vehicle traffic monitoring
Collision and congestion avoidance
Law enforcement
Broadband transmission
Highway lane reservation
Emission control

upon the preceding, one obvious application for an ad hoc network is to provide a vehicular traffic monitoring capability that provides real-time traffic information to drivers. We can consider this to represent information that can be integrated into a vehicle's navigation system to provide the vehicle operator with alternate routing suggestions. In addition, other information could be used by vehicle sensors to automatically control braking when traffic density increases in front of the vehicle. Thus, there is a wide range of applications that could be supported by an ad hoc vehicle network.

Placing our thinking cap back on, we can list a number of specific application areas where information gathered by a Vehicle Ad Hoc Network could be used. Some of those application areas are listed in Table 7.1 and will be discussed in the order listed in the referenced table.

7.1.4.1   Vehicle Traffic Monitoring

By relaying information about traffic flowing through certain intersections, vehicle operators can be notified of potential congested areas prior to arriving at them. Such information can be used as input into an optional vehicle navigation system, enabling vehicles equipped with this system to provide the operator alternate routes, if available, that would bypass all or a portion of a congested area.

A second area concerning vehicle monitoring could include the transmission of certain types of vehicle information from vehicles on one side of a highway to vehicles on the other side of the highway. For example, assume 50 percent of vehicles on one side of the highway have their windshield wipers turned on at a particular point in time, while a few seconds later the percentage increases to 65, and in a few more seconds to over 70. This information could be used to inform approaching traffic that they are entering an area of precipitation. In addition, this information could also be used to automatically turn on the wipers of vehicles approaching the area of precipitation if the vehicle operator has not already done so.

7.1.4.2   Collision and Congestion Avoidance

Logically following vehicle traffic monitoring, information about potential congested areas can be relayed to other vehicle operators, enabling them to consider the use of alternate streets or highways to reach their destination. In addition, because many vehicles are manufactured today so that air bag deployments are transmitted via cell phone to a monitoring center, such as OnStar, it is also within the realm of possibility for such information to be relayed over an ad hoc network. Thus, members of that network could be notified that they are proceeding toward a location where an accident occurred or are in the path of approaching emergency vehicles. Then, via the integration of such information into a vehicle's navigation system, they could be presented with alternate routes that would allow them to bypass the collision area, or could be instructed to pull over to allow emergency vehicles to safely proceed to the area of an accident.

7.1.4.3   Law Enforcement

One of the more modern uses of variable highway signs is to generate AMBER Alerts when a child is missing. Unfortunately, signs may not be visible when drivers are dealing with the navigation of their vehicle through a congested area. In addition, depending upon the route of a vehicle, a fixed location sign may not be available for viewing. Thus, the ability of ad hoc networks to relay AMBER Alerts could represent a significant added capability in the area of law enforcement.

A second area of law enforcement that could be improved upon through the use of ad hoc networking is stolen vehicle recovery. Today the use of several global positioning system (GPS) and cellular notification systems enables the general location of a stolen vehicle to be communicated to law enforcement personnel. In the future, it could be possible to integrate a stolen vehicle notification system into a Vehicle Ad Hoc Network, enabling law enforcement personnel to more rapidly learn that a stolen vehicle is not only at a certain location, but, in addition, is moving toward another location, where a patrol car could be located. Thus, combining the two technologies could enhance the recovery of stolen vehicles, which in turn might reduce the cost of automobile insurance.

7.1.4.4   Broadband Transmission

Because a majority of the ad hoc vehicle networking effort involves the use of wireless LAN technology, we can assume that the development of this type of network will result in vehicles having a broadband transmission capability. This capability could be used by a diverse series of applications. For example, real-time game playing could be performed by young adults (hopefully not while driving) that could enhance the infotainment capability of the vehicle. Other possible uses of the vehicle's broadband capacity could include providing a connection to a nearby hospital in the event of a medical emergency, allowing a person's vital information to be sent to the ER prior to the arrival of the patient.

7.1.4.5 Highway Lane Reservation

One of the more common driving situations many individuals remember is the rapid approach of an emergency vehicle or the sudden stop of a school bus. Both situations result in the modification of a vehicle operator's routine, causing the operator to attempt to either move his vehicle into the slow lane or bring his vehicle to a complete stop.

Through the use of an ad hoc network it becomes possible for designated vehicles, such as police cars, ambulances, and school buses, to transmit messages to either reserve highway lanes for their use or inform vehicle operators of their presence and the need of the driver to take some specific action, such as moving into the slow lane. Thus, the use of an ad hoc networking capability could supplement the audio and visual indicators used by emergency vehicles and school buses.

7.1.4.6 Emission Control

Although this author could probably generate many additional applications that a vehicle's ad hoc network could support, we will conclude our examination of such applications by turning our attention to one that can adversely affect the health of many persons — emission control. When the density of vehicles in an area increases due to an accident or simply too many vehicles attempting to cross an intersection, the density of pollutants from the exhaust of vehicles also increases. This increase in pollutants represents a health hazard to drivers and passengers as well as pedestrians in the area.

Using power management technology incorporated into notebook computers as an example of how different components of the computer can be controlled, it is possible for congestion causing a high concentration of vehicle pollution to be treated in a similar manner. That is, a vehicle's engine behavior could be modified by the exchange of real-time information over an ad hoc network. Behavior modifications could include the adjustment of a vehicle's idling speed or the switching of hybrid vehicles from gasoline to electricity. Now that we have an appreciation for some of the applications that can be supported by vehicle-based ad hoc networks, let us turn our focus to communications technologies that could be used to support the development of such networks, and some of the technical challenges to the development of this type of network.

7.1.5   Communications Technologies

An ad hoc network requires an existing network technology as a transport mechanism. Currently, cellular and satellite-based GPS communications are used in vehicles equipped with GM's OnStar or vehicles that are using a similar telematics service, while Institute of Electrical and Electronics Engineers (IEEE) wireless LAN technology in the form of IEEE 802.11a/b/g standards has been used for experimentation in the development of trial Vehicle Ad Hoc Networks. Thus, in this section we will turn our attention to the IEEE family of wireless LANs, beginning with the basic IEEE 802.11 standard.

7.1.5.1   IEEE 802.11 Standard

The basic IEEE 802.11 standard defined three wireless technologies that could operate at either 1 or 2 Mbps. The three technologies defined by this standard included frequency-hopping spread spectrum (FHSS), direct-sequence spread spectrum (DSSS), and infrared (IR). Although several vendors produced access points and adapter cards that used FHSS and DSSS, to this author's knowledge no products were manufactured using IR.

Although equipment using the basic IEEE 802.11 standard was manufactured by several vendors, the low-data-rate support by the standard severely limited its adoption. It was not until the 802.11a and 802.11b extensions to the standard were developed that the use of wireless LAN technology was increasingly adopted for use in homes and offices.

7.1.5.1.1   IEEE 802.11a Standard

The IEEE 802.11a standard is actually an extension to the 802.11 basic standard, which specifies the use of the 5-GHz frequency band for operation. IEEE 802.11a equipment uses a frequency division multiplexing system, which supports data rates up to 54 Mbps in the 5-GHz frequency band. Specifically, 802.11a equipment can operate in one of eight independent 200-MHz channels from 5.15 to 5.35 GHz. Another 100 MHz of spectrum from 5.725 GHz to 5.825 MHz is available for outdoor use, with a maximum power output of 1 W. Concerning the 200 MHz of frequency that is available for indoor use, the first 100 MHz is restricted to a maximum power output of 50 mW, while operations in the second 100 MHz can occur at 250 mW.

At the physical layer the 802.11a standard specifies the use of orthogonal frequency division multiplexing (ODFM). OFDM uses a series of low-speed subcarriers instead of a single carrier, with each subcarrier orthogonal to one another, which enables small pieces of data to be modulated and transmitted in parallel. Under OFDM, 52 subchannels are used, each approximately 300 kHz wide. Of the 52 subchannels, 48 are used to transmit data, while the remaining 4 are used for error correction.

The 802.11a extension supports several modulation methods. Binary phase shift keying (BPSK) is used to provide 125 kbps of data per subcarrier, resulting in a 6-Mbps data rate. When quadrature phase shift keying (QPSK) is used, the data rate per subcarrier is doubled, resulting in a 12-Mbps data rate. A 16-level QAM can be used to encode four bits per hertz, resulting in a data rate of 24 Mbps, while a 64-level QAM provides the ability to encode eight bits per hertz, which results in a 1.125-Mbps data rate per subcarrier or a 54-Mbps data rate when all 48 subcarriers are used. The 5-GHz frequency band used by the 802.11a extension is unlicensed. Its frequency is almost double that defined for use by other IEEE 802.11 standards. Because frequency is inversely proportioned to transmission distance, the use of an IEEE 802.11a wireless LAN as the basis for a Vehicle Ad Hoc Network would limit the obtainable transmission distance. Thus, most trials have avoided the use of IEEE 802.11a technology due to its limited transmission distance, even though the 5-GHz frequency band used by the technology has a limited amount of interference.

7.1.5.1.2   IEEE 802.11b/g Standards

Both the IEEE 802.11b and 802.11g standards use the 2.4-GHz ISM (industrial, scientific, and medical) frequency band. The use of the 2.4-GHz frequency band provides a transmission range two to four times that obtainable through the use of IEEE 802.11a technology, which utilizes the 5-GHz frequency band.

Initially the majority of field trials involving the use of wireless LANs for the development of a Vehicle Ad Hoc Networking capability involved the use of IEEE 802.11b technology. In actuality, 802.11b represents an extension to the basic 802.11 standard, which was released during 1999 and which raised the transmission rate from 1 and 2 Mbps to 5.5 and 11 Mbps using DSSS, with backward compatibility with the lower data rates. Thus, the IEEE 802.11b standard supports data transmission rates of 11, 5.5, 2, and 1 Mbps. In comparison, the 802.11g specification, which was released during 2003, resulted from the need of some users for the higher data rate of 54 Mbps provided by the IEEE 802.11a standard, but it recognized the importance of using the lower-frequency 2.4-GHz band to maintain a relatively long transmission distance.

Similar to the 802.11a extension, the 802.11g modulation scheme is OFDM. Data rates of 6, 9, 12, 18, 24, 36, 48, and 54 Mbps are supported using OFDM with different modulation techniques, while backward compatibility with 802.11b results in the support of data rates of 5.5 and 11 Mbps. In an outdoor environment, equipment compatible with the IEEE 802.11g standard can normally achieve a transmission range of several hundred meters, which is normally sufficient for a Vehicle Ad Hoc Networking capability. Although ad hoc networking trials were beginning to use IEEE 802.11g technology when this book was researched, another technology is on the horizon that can provide both higher data rates and extended transmission distances, and deserves some mention. That technology is based upon the IEEE 802.11n evolving standard.

7.1.5.1.3   IEEE 802.11n Standard

In January 2004 the IEEE announced that it would commence work on a new standard for wireless networks that would significantly increase the transmission rate and distance of transmission. Designated as the IEEE 802.11n standard, the technology is based upon the use of multiple antennas, which provides a multiple-input, multiple-output (MIMO) transmission capability. As of mid-2007 several vendors had released products based upon preliminary 802.11n specifications; however, the ratification of the standard was more than likely at least a year away.

7.1.5.1.3.1   Overview — The scope of the IEEE 802.11n task group developing the standard is to define modifications to the wireless LAN physical layer and Medium Access Control layer (PHY/MAC) that provide a minimum 100-Mbps data transfer capability. This minimum throughput represents an approximate fourfold increase in wireless LAN transmission when compared to the IEEE 802.11g standard. Table 7.2 compares the data transmission rates and frequency band use of the alphabet soup of IEEE wireless LAN transmission standards.

The key to the ability of the evolving IEEE 802.11n standard to increase the physical data transfer rate is the use of MIMO technology. MIMO exploits the use of multiple signals to enhance wireless performance. To do so, a MIMO-capable receiver needs to simultaneously process spatially different signals, while a MIMO-capable transmitter must simultaneously transmit multiple signals. Thus, the use of MIMO requires IEEE 802.11n equipment to support both antenna diversity and spatial multiplexing.

Table 7.2 Wireless LAN Maximum Transmission Rates and Frequency Band Use

IEEE Standard	Maximum Transmission Rate (Mbps)	Frequency Band Use (GHz)
802.11	2	2.4
802.11a	54	5.0
802.11b	11	2.4
802.11g	54	2.4
802.11n	200	2.4/5

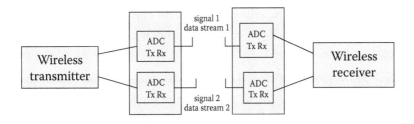

Legend: ADC analog digital converter

Figure 7.3   A two-antenna MIMO system.

7.1.5.1.3.2   Multiple Antennas — MIMO requires the use of multiple antennas at both the transmitter and receiver. Figure 7.3 illustrates a two-antenna MIMO system, which results in transmission via a two-stream spatial division multiplexing (SDM) arrangement.

In examining Figure 7.3, note that the use of multiple antennas facilitates the ability of the receiver to resolve information flowing on multiple signal paths. The actual signals transmitted are referred to as multipath signals because they bounce off different objects and form multiple paths to the receiver. Although multipath signals are normally thought of as interference that degrades the ability of a receiver to recover information, under MIMO such signals are spatially resolved, which actually enhances the ability of the receiver to recover information.

7.1.5.1.3.3   Spatial Division Multiplexing — The second key to the ability of MIMO technology to significantly increase data transmission rates is its use of spatial division multiplexing (SDM). Through the use of SDM, multiple independent data streams can be simultaneously transmitted within one spectral channel of bandwidth, minimizing frequency requirements. To do so, each spatial data stream requires its own transmit/receive antenna pair at each end of the transmission, as illustrated in Figure 7.3.

7.1.5.1.3.4   Compatibility — Due to the large base of IEEE 802.11a/b/g devices that have reached the market, the task group developing the 802.11n standard requires backward compatibility with the earlier standards. Thus, the evolving IEEE 802.11n standard will not only provide an enhanced data transmission and transmission range, but also enable communications with older wireless LAN standards.

Perhaps the key advantage to the use of 802.11n technology as a basis for a Vehicle Ad Hoc Network is its transmission range. In an outdoor environment it is possible to have a transmission distance of approximately 500 ft, or about 1/10 of a mile. This transmission range considerably exceeds the range of earlier IEEE802.11

standards and would enable nodes in a Vehicle Ad Hoc Network to remain as part of the network longer than if earlier wireless technology was used.

7.1.5.1.3.5   Summary — Due to the miniaturization of antennas, the requirement for multiple antennas in a vehicle to support the IEEE 802.11n standard should not be difficult. Because this new standard provides true broadband communications while being backward compatible with prior wireless LAN standards, it provides a communications technology that can support previously mentioned ad hoc networking applications as well as many additional applications. For example, many municipalities are either constructing or developing municipalwide wireless LAN coverage by creating a mesh network based upon older wireless LAN technology. Thus, a vehicle that uses wireless LAN technology for ad hoc networking could also access the Internet as the vehicle transits WiFi hot spots to retrieve e-mail, surf the Web, search for a hotel or eatery, and perform other types of Internet-based activities.

7.1.6   Vehicle Frequency Utilization

Ad hoc networking needs to be examined with respect to the frequency use of other applications that are incorporated into vehicles. Thus, in this section we will discuss the operational frequency of wireless applications integrated into vehicles.

Table 7.3 lists wireless technology standards and applications that can be expected to be supported by future vehicles. Although it is not certain that each vehicle will support every entry in Table 7.3, due to the use of different frequencies by various applications and standards, it is imperative that evolving applications use standardized frequencies to minimize or avoid frequency interference issues.

Table 7.3   Existing and Emerging Wireless Technology's Applications

AM radio
Bluetooth
FM radio
GPS navigation
Satellite radio
Short-range radar
Wireless LAN

7.1.6.1   AM Radio

Analog modulation (AM) radio occurs in the medium waveband. Specifically, frequencies from 530 to 1710 kHz are allocated to AM radio broadcasts.

7.1.6.2   Bluetooth

As mentioned earlier in this book, Bluetooth represents a short-range, low-power wireless transmission technology. Bluetooth operates in the 2.4-GHz frequency band, which is also referred to as the industrial, scientific, and medical (ISM) unlicensed frequency band. Although Bluetooth operates in the same frequency band as microwave ovens, most wireless LANs, and portable phones, its use of frequency hopping commonly minimizes interference between Bluetooth devices and other wireless devices operating in the same frequency band.

7.1.6.3   FM Radio

Frequency modulated (FM) radio occurs in the 88- to 198-MHz frequency band, commonly referred to as very high frequency (VHF).

7.1.6.4   GPS

GPS represents a low-power navigation signal that is commonly used in DVD-based navigation systems to show the location of a vehicle with respect to the display of a portion of a map of the area the vehicle is traversing. Due to the low power of GPS, it is a line-of-sight communications method that does not work if a vehicle is located in a garage or traversing a tunnel. GPS signals occur at 1227.6 and 1575.42 MHz.

7.1.6.5   Satellite Radio

One recent addition to the use of the frequency spectrum is satellite radio. Technically referred to as Digital Audio Radio Service (DARS), satellite radio is familiar to many new vehicle purchasers, who receive a free six-month or year subscription with their new vehicle. There are two major satellite radio operators, Sirius Satellite Radio and XM Satellite Radio, with both operators using the 2.3-GHz S-band. XM Radio uses two geostationary satellites and transmits using the 2332.50- through 2345.00-MHz frequency band. Sirius Satellite Radio uses three satellites and broadcasts on designated frequencies between 2320.00 and 2332.50 MHz.

7.1.6.6   Short-Range Radar

Over the past decade, a number of frequency bands have been used for inter-vehicle radar operations. In addition, several frequency bands have been allocated to the use of radar detectors. In this section, we will first review the use of frequencies for radar detectors, and then turn our attention to the frequencies used for various field trials involving the use of radar as a safety mechanism for inter-vehicle communications.

7.1.6.6.1   Radar Detectors

K-band radar guns date to 1978 and operate in the 24.05- to 24.25-GHz frequency band. K-band guns keep the transmitter in a hot standby mode of operation, which enables an officer to activate the gun when a vehicle is within 200 to 300 yards. The initial K-band handheld radar guns could only operate from a stationary position; however, a pulsed version was introduced that enables the gun to be used from both a stationary position or a moving vehicle.

In 1987 the U.S. Federal Communications Commission (FCC) allocated the Ka-band for police radar use. Various radar guns were developed that operate in the Ka-band from 34.2 to 36 GHz.

A third frequency band allocated for police radar is the X-band. In actuality, this was the first frequency band allocated for police radar, and it dates to the 1950s. X-band radar operates in the 10.5- to 10.55-GHz frequency band.

7.1.6.6.2   Radar as a Safety Mechanism

There are several frequency bands currently used in field trials to enhance the ability of vehicles to recognize other vehicles they are approaching or vehicles that are approaching them. Typically such field trials use radar to determine when a vehicle should begin braking or issue an audio or visual warning in the event a driver attempts a lane change while another vehicle approaches that is obscured from view. The radar frequency band for such field trials varies based upon location, cost, and other variables. However, the location of the field trial represents a major consideration because the allocation of frequencies for different applications is performed by various administrative bodies, such as the FCC in the United States and the European Union (EU) Commission in Western Europe. Table 7.4 lists three radar frequency bands currently used in field trials. In addition to those bands, the 60-GHz band is being considered for use in the United States, and the 79-GHz band is being considered for use in Western Europe. The latter is being considered because other radio services are already using the 24-GHz band, which extends from 21.65 to 26.65 GHz. Thus, the use of the 24-GHz frequency band could result in interference with radio applications that are already being used.

Table 7.4   Examples of Radar Used for Vehicle Safety Trials

Frequency (GHz)	Description
24	Allocated by the EU Commission for Driver Assistant Systems (DAS)
35	FM continuous wave used for detecting intruders and moving vehicles, traffic monitoring and control
76.5	FM continuous wave prototype radar used for automobile collision warning, traffic monitoring and detection

7.1.6.7   Wireless LANs

Both the 2- and 5-GHz frequency bands are used for wireless LAN operations. Each of these frequency bands represents areas in the frequency spectrum that can be used without having to obtain a license, hence the term *unlicensed frequency band*. However, it is important to note that although wireless LANs operate in unlicensed frequency bands, this does not mean that their use is not regulated. In the United States the FCC regulates the maximum power that can be used by a transmitter in each frequency band. For example, IEEE 802.11b- and 802.11g-compatible devices can have a maximum transmission power of 100 mW.

7.1.6.7.1   Protocols

A Vehicle Ad Hoc Network represents a Mobile Wireless Ad Hoc Network. Thus, the term *MANET* is commonly used to refer to this type of network, even though its early evolution was oriented toward the use of portable computers in trains and automobiles. Since the late 1990s, a considerable amount of academic papers have been published concerning the use of different routing protocols in a MANET.

7.1.6.7.2   MANET vs. VANET

Although a MANET can be considered to represent a Vehicular Ad Hoc Network (VANET), there is a key difference between the two that can make some MANET models obsolete when applied to a VANET. That difference is in the movement of traffic. Instead of using random movements, a VANET results in vehicles moving in an organized flow because vehicles are restricted in their range of motion, as they rationally need to flow on streets and highways according to traffic laws.

7.1.6.7.3   MANET Protocols

In this section we will focus our attention on several MANET protocols, as they will more than likely form the basis for developing a future VANET protocol standard.

One protocol that shows promise to be adapted to use in a VANET is the Ad Hoc On-Demand Distance Vector (AODV) routing protocol, while a second protocol known as Topology Dissemination Based on Reverse-Path Forward (TBRPF) also could be used in modified form as the routing protocol for VANETs.

7.1.6.7.3.1 Types of Protocols — Most MANET protocols can be categorized as either being proactive or representing an on-demand or reactive type of protocol. Protocols that can be categorized as proactive update the routing information by exchanging route data at periodic intervals. Such exchanged route data is placed into tables in each device and provides information on routing prior to devices requiring route data. Thus, a proactive routing protocol represents a mechanism to reduce network latency because there is no need to determine a route when data needs to be transmitted. However, because of the periodic updating of route tables regardless of whether such data is needed, a proactive network routing protocol can have a relatively high overhead.

A second category of MANET routing protocols performs route maintenance only when information needs to flow on a new route. This type of protocol is reactive as it responds to the need to determine a route. Another moniker or name for this type of reactive protocol is "on demand," because a route is determined only when needed. Because the exchange of routing information occurs when needed, the overhead associated with an on-demand routing protocol is typically less than that for a proactive routing protocol. However, because there is no free lunch in communications, the lower overhead occurs at the expense of an increase in latency, which results from devices having to learn routes at the time information needs to be transferred. Because of the emergence of the AODV routing protocol as popular for use in wireless mesh networks, we will focus our attention upon how this protocol operates. Once this is accomplished, we will look at a second protocol: TBRPF, which represents an experimental, proactive routing protocol currently used by a few wireless mesh networking vendors.

The AODV Routing Protocol — As a review, when two or more devices come within close proximity of one another and require the ability to exchange data, we have a special type of network. That type of network is referred to as an ad hoc network. This type of network can be vehicles equipped with a common wireless LAN infrastructure that come within range of one another.

Need for a Routing Protocol — When only two vehicles are participants in an ad hoc network and in range of one another, they can directly exchange information. Thus, there is no need for a routing protocol. However, when three or more vehicles come together in an ad hoc network a routing capability can become a necessity. To see why this is true, consider Figure 7.4, which illustrates an ad hoc network consisting of three vehicles that represent network nodes. If we assume vehicles A and C are not within range of each other, then the only way they can reach each other is via a route through vehicle B, because B's service range overlaps A and C. Thus,

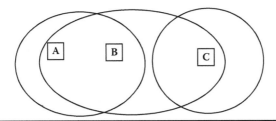

Figure 7.4   An ad hoc network consisting of three vehicles using wireless technology.

vehicle B must learn how to access A and C, while vehicles A and C must learn that the route to each other is through vehicle B. Although routing in a Vehicle Ad Hoc Network with a larger number of vehicles becomes more complicated, Figure 7.4 illustrates the need for a protocol that allows computers to use the facilities of other devices to access distant vehicles.

AODV — Based upon the requirement to route messages between vehicles, several routing protocols were developed for use by Mobile Ad Hoc Networks. In this section, we will examine the messages used to route data under the AODV routing protocol and the manner by which the protocol operates. Because AODV can be tailored to operate with vehicles, it provides a basis for understanding how a Vehicle Ad Hoc Networking protocol could operate.

Route Request Message — In an AODV environment routes between vehicles are created only as needed, hence the term *on demand* is used to reference this protocol. Once a route is required to a particular destination, the AODV protocol transmits a Route Request (RREQ) message packet, which propagates across the wireless network. The format of the RREQ message is shown in Figure 7.5.

In examining the format of the RREQ message packet, note the use of destination and origination sequence numbers. The destination sequence number is created by the destination and is included with route information it transmits to requesting nodes. If two routes are defined to a destination, the requesting node is required to select the one with the highest sequence number. Here the higher sequence number indicates a newer or fresher route.

In addition to providing a mechanism to select a route when more than one is available, sequence numbers enable AODV to avoid routing loops that could result in messages repeatedly propagating over the same path. To ensure sequence numbers are updated under AODV, the broadcast of RREQ messages as well as RREP and RERR messages, which we will examine shortly, includes sequence number fields that are incremented prior to transmission.

Each node that receives a RREQ message packet checks the destination IP address field to determine if the node represents the destination. If the node is not the destination and does not have a route to the destination, it rebroadcasts the RREQ message to its immediate neighbors as well as updates its route table by including a reverse pointer to the originator. This process continues until a route to

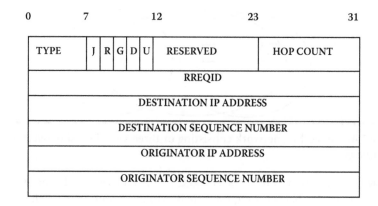

Legend

J Join flag, reserved for multicasting
R Repair flag, reserved for multicast
G Gratuitous flag, indicates if a gratuitous RREP should be unicast
   to the node specified in the destination IP address field
D Destination only flag, indicates only the destination should
   respond to this RREQ message
U Unknown sequence number, indicates destination number unknown

Figure 7.5 Route Request message format.

the destination node is located or the IP datagram transporting the RREQ message reaches its maximum hop count and is discarded.

Route Reply Message — When the RREQ message packet either reaches the destination node or encounters a node with a route to the destination, a response is transmitted. That response occurs via the transmission of a Route Reply (RREP) message. The RREP message packet, whose format is shown in Figure 7.6, flows toward the originating or source node. As the RREP message packet flows through intermediate nodes, such nodes update their route information about source and destination nodes.

In examining Figure 7.6, note that setting the A bit field requires a Route Reply Acknowledgment (RREP-ACK) message to be returned. Two other fields in the RREP message differ from the RREQ message and also deserve mention: the prefix size and lifetime fields.

The five-bit prefix size field, when set to a nonzero value, specifies that the indicated next hop can be used for any nodes with the same routing prefix as the requested destination. Thus, the prefix size field enables a subnet router to provide a route for every host in the subnet defined by the routing prefix. In comparison, the 32-bit lifetime field contains the time expressed in milliseconds for which nodes receiving the RREP message packet consider the route to be valid.

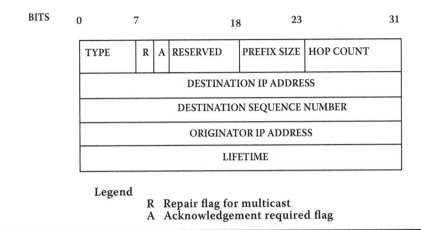

Legend
R   Repair flag for multicast
A   Acknowledgement required flag

Figure 7.6   Route Reply (RREP) message format.

RREQ-RREP Message Flow — To illustrate the flow of RREQ and RREP messages we need a network. Thus, let us assume we have a five-node network, as illustrated in Figure 7.7, where node 1 transmits a Route Request (RREQ) message packet in an attempt to determine a route to node 5, the destination vehicle in the network. Because node 2 represents an intermediate node that is not the final destination, this node relays the RREQ message by rebroadcasting it to nodes 3 and 4.

As the RREQ message continues its propagation through the network it reaches node 3. At that node the vehicle determines it is not the destination, nor does it have a path to another node. Thus, the message packet is dropped by node 3. When the RREQ message packet reaches node 4, that node forwards the packet to node 5, which is the destination node. While the RREQ message propagates through the network, each intermediate node observes the value of the packet's originator IP

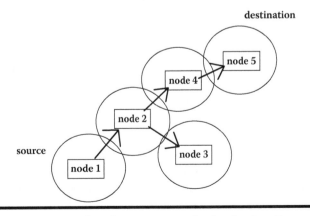

Figure 7.7   RREQ message flow from source to destination in a five-node network.

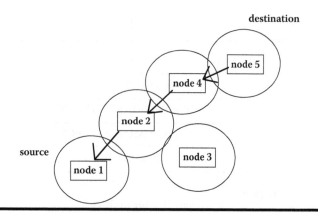

Figure 7.8 RREP message flow in response to a prior RREQ message.

address field, enabling the route information table for the source node to be updated at each node, using the neighbor that propagated the packet as the next hop.

Once the destination node receives the RREQ message it will respond with a RREP message packet. The RREP message packet issued by the destination node, which in our small example is node 5, flows to node 4. At node 4 the computer lookup of its route table indicates that the next hop toward the source node address is node 2. Thus, the RREP is propagated to node 2. At that node another table lookup occurs and node 2 notes that the source node is node 1 and relays the RREP to node 1. This sequence is illustrated in Figure 7.8.

In the preceding example of the determination of a route from source to destination, the RREQ message packet was shown flowing to the destination. In actuality, a route can be determined when the RREQ reaches either the destination or an intermediate node that has a route to the destination. That route should be a fresh route, which means that the route is a valid entry for the destination, whose associated sequence number is at least as great as the sequence number contained in the RREQ message. Now that we have an appreciation for the manner by which routes are determined under AODV, let us turn our attention to their deletion.

Route Deletion — Once a route is determined it will remain in each vehicle's node routing table until the route becomes inactive for a period. Each routing table entry contains an active route timeout field that is updated each time the route is used to forward a data packet. If the timeout value expires, the route is then deleted.

A second method can also result in the deletion of a route. That method involves the failure of a node to receive Hello messages. Thus, let us briefly focus our attention on this message.

Hello Message — A node can provide connectivity information by broadcasting Hello messages when it is part of an active route. A Hello message is a RREP message with a time-to-live (TTL) value of 1 set in the IP header of the message. In the RREP the destination IP address field is set to the node's IP address that is

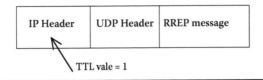

IP Header	UDP Header	RREP message

TTL vale = 1

Figure 7.9   An encapsulated AODV Hello message.

broadcasting the Hello message. The destination sequence number field is set to the node's latest sequence number, while the hop count field is set to a value of 0.

If you examine the format of the Route Request message illustrated in Figure 7.5, you will not see a TTL field. The reason for this is due to the fact that AODV messages are transported via the User Datagram Protocol (UDP). UDP messages are in turn prefixed with an Internet Protocol (IP) header to form an IP datagram, with the IP header containing the TTL field, whose value is then set to 1 when an RREP message is used as a Hello. Figure 7.9 illustrates the formation of an IP datagram that becomes an AODV Hello message.

If a node fails to hear a predefined number of consecutive Hello messages from its next-hop neighbor, the route table entry will be deleted by the listening node. When this action occurs, it represents a break in network connectivity, which results in the transmission of a Route Error (RERR) message packet. That packet is then propagated back to the source node to inform all nodes that have that route in their routing table that the route is unusable and should be deleted from their routing tables. When a destination becomes unreachable, a node will transmit a Route Error message. Thus, let us turn our attention to the format and operation of this message.

Route Error Message — Figure 7.10 illustrates the format of the Route Error (RERR) message. Note that the destination count field indicates the number of unreachable destinations included in the message. This field must have a value of at least 1. Similar to other AODV messages, the RERR message is transported via UDP within an IP datagram.

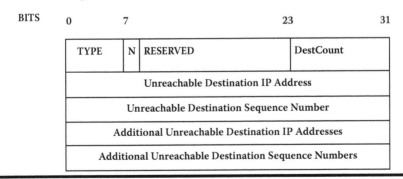

| BITS | 0 | 7 | | 23 | 31 |

TYPE	N	RESERVED	DestCount
Unreachable Destination IP Address			
Unreachable Destination Sequence Number			
Additional Unreachable Destination IP Addresses			
Additional Unreachable Destination Sequence Numbers			

Figure 7.10   Route Error message format.

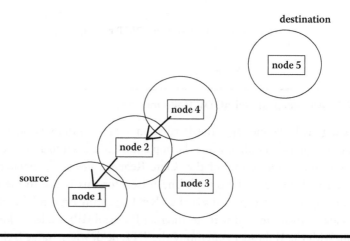

Figure 7.11   RERR message flow assuming node 5 becomes detached from the network.

To illustrate the flow of the RERR message, let us assume node 5, shown in Figure 7.7 and Figure 7.8, fails or moves outside the range of node 4. When this situation occurs, node 4 will not hear Hello messages from node 5. After a predefined number of Hello messages are missed, node 4 deletes the route to node 5 from its routing table and transmits an RERR message that marks the route to node 5 as invalid. The RERR message is transmitted to neighbor nodes that were using node 4 as the next hop for the route to node 5. Thus, the RERR message is transmitted to node 2. Similarly, node 2 transmits the RERR message to node 1. After receiving the RERR message, the computer at each node deletes the route to the unreachable node from its routing table. If a route to the destination that was just deleted becomes required, the source node that needs the route will initiate a new route discovery process. Figure 7.11 illustrates the flow of RERR message packets for our five-node network, assuming node 5 became separated from the network, as the vehicle that represents that node accelerates so that it is not within range of the cluster of vehicles represented by nodes 1 through 4.

In concluding our discussion of the AODV routing protocol we will turn our attention to a situation referred to as a gray area. In doing so we will note how one vendor modified the AODV protocol to overcome this situation.

Gray Area Considerations — A gray area occurs when the wireless signal between a mobile node and a fixed location becomes too weak for application data to reach its destination. To illustrate how a gray area can occur and the manner by which one vendor modified the AODV routing protocol, we need a sample network for illustrative purposes. Thus, let us create one.

Figure 7.12 illustrates two wireless networks interconnected via a pair of wireless bridges. Each network uses the AODV routing protocol to support mobile ad hoc operations, even though the bridges interconnecting the two separated

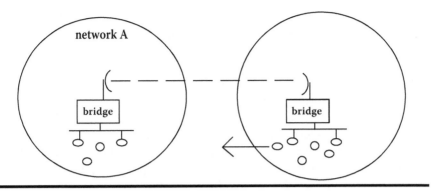

Figure 7.12 The addition of signal strength to control packets can be used to minimize the effect of gray zones.

geographical areas represent stationary devices. If we assume that just one host is mobile, let us examine what happens as it moves from the area served by bridge A to the area served by bridge B.

When the mobile node is within a relatively short distance of other nodes in the network supported by bridge A, it can receive a high enough level of signal strength to become a participant in the mesh network at that location. However, as the mobile node moves from the area serviced by bridge A to the area serviced by bridge B, the signal strength between the mobile node and other wireless nodes that make up network A decreases. At a certain distance between networks A and B the mobile node's ability to pick up signals from the wireless nodes in network A degrades to the point where the signal strength is just strong enough for periodic control messages to link the mobile node to network A. In this situation a route from a node in network A to the mobile node may remain in the routing tables of other nodes in network A even though the actual data transfer capability is diminishing toward zero. In fact, as long as one node in network A continues to receive a minimum number of Hello messages from the mobile node, a route to the mobile node will continue to occur via network A. Only after the mobile node is completely outside the range of other nodes in network A will a node in network B become an intermediate node for the route to the mobile device. The area from which the mobile device loses its ability to communicate with nodes in the network it is leaving until it joins the new network is commonly referred to as a gray area or gray zone.

One interesting method used to counter the gray zone problem involves a change to the manner by which the AODV protocol operates. Under the implementation of the AODV routing protocol by Nova Roam, each control packet includes a signal strength metric. If the control packet signal strength metric falls below a user-defined threshold, an existing link between nodes will be dropped, in effect forcing the affected nodes to find new routes. If we return to the example shown in Figure 7.12, employing the modification to AODV used by Nova Roam, as the mobile node moves toward network B its signal strength to the nearest node in network A diminishes to the point where its connection is dropped. As the mobile node reestablishes a route, it moves closer to a node in network B and finds a new route to network B. Thus, the addition of a signal strength indicator can be employed to minimize the gray zone or gray area problem that occurs when truly mobile nodes move between areas where a grouping of nodes forms a wireless mesh network.

TBRPF — The Topology Dissemination Based on Reverse-Path Forwarding (TBRPF) routing protocol, a mouthful to utter, is an experimental protocol adapted by a few wireless mesh networking equipment vendors in modified form. TBRPF is defined in RFC 3684 and can be considered to represent a proactive, link-state routing protocol developed for use in Mobile Ad Hoc Networks.

Overview — TBRPF is a proactive link-state routing protocol that provides hop-by-hop routing along the shortest paths to each destination in a network. Each node in a network using TBRPF computes a source tree, which consists of paths to all reachable nodes based upon practical topological information stored in a topology table in the nodes. To minimize transmission overhead, each node transmits only a portion of its source tree to neighboring nodes, using a combination of periodic and differential updates to keep its neighbors informed of the reported portion of its source tree. Here the differential updates report only changes in the status of neighbors. Each node can optionally report additional topology information with a resulting increase in overhead. In fact, it is possible to configure nodes in a robust mobile network environment using TBRPF as the routing protocol to transmit its full topology to neighbors.

Protocol Modules — The TBRPF protocol consists of two main modules. One module is responsible for Neighbor Discovery and is known as the Neighbor Discovery module, while the second module performs topology discovery and route computations. The later is referred to as the routing module.

Neighbor Discovery Module — The TBRPF Neighbor Discovery (TND) module is responsible for discovering neighbors in the network. This protocol enables each node (i) to quickly detect neighbor nodes (j), such that a bidirectional link (I, J) exists between an interface I of node i and an interface J of node j.

Similar to the AODV protocol, the TBRPF Neighbor Discovery module uses Hello messages. However, those messages can be differential in that they only report changes in the status of a link. For example, differential Hello messages would include only the IDs of new neighbors and recently lost neighbors, instead of information about all neighbors. This action can result in Hello messages that are significantly smaller than those of other link-state routing protocols. This enables Hello messages to be transmitted more frequently and allows faster detection of topology changes that can be an important consideration if nodes are truly mobile.

Because TND is designed to be fully modular and independent of the routing module, it only performs direct neighbor sensing. That is, it determines nodes that are one-hop neighbors, resulting in the routing module being responsible for discovering neighbors at a greater distance.

If a node has multiple interfaces, TND is separately run on each interface. This action results in the construction and maintenance of a neighbor table for each local interface. The neighbor table is responsible for storing the status of each link, such as 1-Way, 2-Way, or Lost. The contents of Hello messages are then used to update the contents of the neighbor table, while the contents of the neighbor table determine the contents of Hello messages.

The actual specifications for the operation of TND is well thought out, as it includes methods to ensure a node will not inadvertently miss the declaration of a link being lost, nor will it establish a link that will be short-lived. Concerning the former, when a node changes the status of a link, it will commonly issue three consecutive Hello messages. The node at the opposite end of the link will either receive one of the Hello messages or miss all messages, with the latter situation causing the node to declare the link lost. Concerning short-lived links, nodes must receive a specified number of Hellos prior to declaring the link to be operational. In this manner the counting of Hello messages becomes an important criterion for acquiring or losing a link.

Routing Module — Each node operating TBRPF maintains a source tree that indicates the shortest paths to all reachable nodes in the network. Using partial topology information stored in its topology table, each node uses a modified Dijkstra algorithm to compute its topology table. As a refresher, the Dijkstra algorithm, which is named after its discoverer, E.W. Dijkstra, solves the problem of locating the shortest path from a point in a graph, referred to as the source, to a specific destination. Because it is possible to find the shortest paths from a given source to all points in a graph at the same time, the problem solved by the Dijkstra algorithm is also referred to as the single-source shortest paths problem.

The Reported Subtree — The portion of the source tree that a node reports to its neighbors is referred to as the *reported subtree*. Thus, if T represents the source tree maintained by each node, then RT becomes the reported subtree. Each node reports RT to its neighbors using a periodic topology update, while additions and deletions occur as differential updates on a more frequent basis. For example, periodic updates could occur every 4 or 5 s, while differential updates could occur every second. The reason periodic updates would not occur at much longer time intervals results from the speed of vehicles. For example, at 70 mph a vehicle travels 1.167 miles/min, which is equivalent to 102.67 ft/s. Thus, in approximately 5 s a vehicle traveling at 70 mph would be out of transmission range of a vehicle exiting a highway, while a vehicle traveling 75 mph passing a vehicle traveling 65 mph would be out of transmission range of the lower-speed vehicle in approximately 34 s.

Through the use of periodic updates new neighbors are informed of the RT, while differential updates are used to rapidly disseminate changes to all nodes affected by the update. The reported subtree (RT) consists of links that include neighbor nodes only when such nodes represent the shortest path to a neighbor. In making this determination, a node computes the shortest path from each neighbor to each other neighbor for up to a two-hop distance, using only neighbors as an

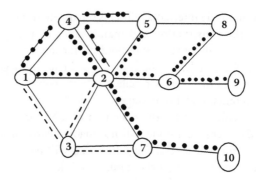

Legend:
●●● node 2's reported subtree
--- node 3's reported subtree
◆●◆ node 4's reported subtree

Figure 7.13   TBRPF networking example.

intermediate node. Although the actual operation of the routing module can be quite complex as the number of nodes in the network increase, we can obtain an appreciation for its basic operation by considering the partial network illustrated in Figure 7.13.

In the example shown in Figure 7.13 let us assume that node 2 selected itself as a parent for all neighbors due to its small ID. As a result, node 2 reports its entire source tree, which in effect is a view of the network without any closed loops. In comparison, nodes 3 and 4 report their one-hop neighbors, which represent a small portion of their trees.

TBRPF Packets — TBRPF is a packet-oriented protocol, with each packet consisting of a header, optional header extension, and body. The latter consists of one or more messages that are filled at the end with padding options that may be necessary for alignment on natural boundaries. The format of the TBRPF packet header is illustrated in Figure 7.14.

In examining the TBRPF packet header, note that four bits define the TBRPF version number, with version 4 of the protocol defined when this book was written. The first two flag bits (L, I) specify which header extensions, if any, follow, while the last two flag bits are presently reserved and are set to a value of 0. The L bit, when set, indicates a 16-bit-length field, which indicates the length of the TBRPF packet. When the I flag bit is set, it indicates that the router identification (RID) field consisting of 4 bytes is contained in the TBRPF packet.

Packet body. The TBRPF packet body follows the header. The packet body consists of the concatenation of one or more TBRPF messages and terminates with optional padding to satisfy any alignment requirements.

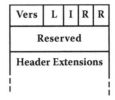

Legend:

Vers Version
L Length Included
I Router ID Included
R Reserved

Figure 7.14 TBRPF packet header format.

Messages. Under the TBRPF protocol there are two major categories of messages that flow between nodes. The first category of messages involves three types of Hello messages that are used for Neighbor Discovery purposes, while the second category of messages involves topology updates.

The basic format of the Hello message is shown in Figure 7.15.

The four-bit type field currently defines three Hello messages. A type field value of 2 defines a Neighbor Request Hello, while type field values of 3 and 4 define Neighbor Reply and Neighbor Lost Hello messages, respectively. The four-bit priority (PRI) field indicates the sending node's relay priority, which is expressed as an integer value between 0 and 15. A node with a higher relay priority is more likely to be selected as the next hop than a node with a lower value. The value 0 is reserved for nonrelay nodes, such as nodes that never forward packets originating from other nodes.

A router in normal operation has a relay priority value of 7 and can dynamically change its relay priority value. The priority field is followed by a 12-bit number

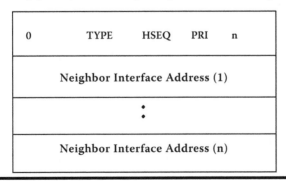

Figure 7.15 TBRPF Hello message format.

Table 7.5   TBRPF Neighbor Table Entries

Entry	Description
Router ID	The router ID of the node associated with the neighbor interface
Status	The current status of the link (Lost, 1-Way, or 2-Way)
Life	Amount of time in seconds remaining before the status must be changed to Lost if no further Hello message is received
Sequence	The value of the sequence number in the last Hello message received on interface I from interface J
Count	The remaining number of times a Neighbor Request, Reply, or Lost message containing J must be transmitted on interface I
History	A list of the sequence numbers of the last HELLO_ACQUIRE_ WINDOW Hello message received on interface I from interface j
Metric	An optional measure of the quality of the link I, J represented by an integer between 1 and 255, in which smaller values indicate better quality
Priority	The relay priority of the node associated with interface J

that indicates the number of 32-bit IPv4 interface addresses in the message. Each of those addresses represents neighbors of the node.

Neighbor Table — As previously discussed, each node maintains a neighbor table for each of its local interfaces. Entries in the neighbor table are obtained from Hello messages received by an interface. Table 7.5 lists the possible entries in the neighbor table for interface I for its neighbor J.

In examining the entries in Table 7.5, note that the value of the Hello sequence number is used to determine the number of Hello messages that have been missed. Also note that an entry for interface J in the neighbor table for interface I may be deleted if no Hello was received on interface I from interface J within twice the last hold time. The hold time, which is expressed in seconds, is computed while Neighbor Lost messages containing J are transmitted. In addition, the absence of an entry for a given interface J is equivalent to an entry with a status value of Lost and a history value of Null.

Status Values — As indicated in Table 7.5, the status value for interface I to J can have one of three values: Lost, 1-Way, or 2-Way. A value of Lost indicates that interface I has not received a sufficient number of Hello messages from interface J. In comparison, 1-Way indicates that interface I received a sufficient number of Hello messages from interface J but the link is not 2-Way. The third possible value, 2-Way, indicates that interfaces I and J both received a sufficient number of Hello messages from each other.

Transmitting Hello Messages — Under the TBRPF protocol each node must transmit on each of its local interfaces at least one Hello message per HELLO_INTERVAL, whose default value is 1 s. Hello messages can be transmitted at more frequent

Table 7.6 Hello Message Content Determination

Neighbor Status	Table Entries Count	Hello Message Content
Lost	>0	Include J in Neighbor Lost message and decrement count
1-Way	>0	Include J in Neighbor Request message and decrement count
2-Way	>0	Include J in Neighbor Reply message and decrement count

intervals; however, the time between two consecutive Hellos on a given interface must be greater than the NBR_HOLD_TIME parameter divided by 128. This avoids the possibility that the Hello sequence number wraps around to the same value prior to a neighbor that stops receiving Hello messages changing the status of the link to Lost. The default value of the NBR_HOLD_TIME parameter is 3 s.

Because the synchronization of control messages can result in collisions, Hello messages should not be transmitted at equal intervals. Instead, nodes select an interval randomly with a value up to the HELLO_INTERVAL value. Each Hello message always includes a Neighbor Request message, where the latter includes the sequence number, which is incremented by 1 (module 256) each time a Hello is sent.

The Hello message will also include a Neighbor Reply message if its list of neighbor addresses is not empty, while a Neighbor Lost message is included if its list of neighbor addresses is not empty. The actual contents of the Neighbor Request, Neighbor Reply, and Neighbor Lost messages depend upon the setting of the status and count entries in the neighbor table. Table 7.6 indicates how the contents of the three messages are determined based upon the values of the previously mentioned neighbor table entries.

An exception to the operations listed in Table 7.6 occurs when a node restarts. When this condition occurs, all entries are purged from the neighbor table. In addition, the node ensures that for each interface at least one of the following two conditions is met:

- The difference between the transmission times of the first Hello sent after restarting and the last Hello before restarting is at least twice the Number_Hold_Time in seconds.
- Letting the parameter value of HSEQ_LAST denote the sequence number of the last Hello message transmitted prior to restarting, the sequence number of the first Hello message transmitted after restarting is set to HSEQ_LAST+NBR-HOLD_COUNT+1 (module 256).

The purpose of the above two conditions is to ensure that when a node restarts, each neighbor that has a link to its interfaces will set the status of the link to Lost.

Processing Hello Messages — When a node receives a Hello message, it performs a series of predefined functions. First, it obtains the IP address of the originating interface from the IP header prefixed to the message. Next, it looks for the router ID field in the TBRPF packet header of the received Hello message. If the Hello message contains a router ID field, the node uses that value; otherwise, the node assigns the router ID equal to the IP address it previously obtained.

In addition to the previously mentioned operations, a node performs other steps, depending upon the current status of the link, the router ID value in the received Hello message, and the presence of an entry for the interface. Such operations are described in detail in RFC 3684, which was issued in February 2004 for the experimental version of the TBRPF protocol. In addition, the referenced RFC lists the parameters used by the Neighbor Discovery protocol portion of the TBRPF protocol, to include their proposed default values, the data structure of the topology tables maintained by each node, and the format of the topology update message. Because the RFC is experimental and the operation of the TBRPF protocol has a good probability of changing, we will conclude this section without going into the explicit details of the protocol's operation. Instead, we will conclude with an overview comparison of the AODV and TBRPF protocols.

Protocol Comparison — There are several features most network managers and LAN administrators need to consider when comparing AODV with TBRPF. Those features include the maximum number of nodes each protocol can support, the ability of the protocol to support multiple routes, and unidirectional and multicast support. In addition, the ability of each protocol to be used in low- and high-mobility scenarios is important and must be considered.

Table 7.7 provides a general comparison of the previously mentioned AODV and TBRPF features. In addition to the four features listed in the table, its important to note that AODV is a reactive protocol, whereas TBRPF represents a proactive routing protocol.

Of the four features listed in Table 7.7, this author would consider the maximum number of nodes supported to be the most critical, if you are in doubt about the potential number of vehicles a wireless mesh network will support.

Traffic Support — When considering the ability of a routing protocol to support traffic, several metrics need to be considered. Those metrics include the ratio between the amount of incoming and actually received data packets (packet delivery ratio),

Table 7.7   Comparing Protocol Features

Feature	AODV	TBRPF
Maximum number of nodes	1000±	200±
Multiple routes	No	Possible
Unidirectional link support	Possible	No
Multicast support	Possible	No

Table 7.8 AODV versus TBRPF

		Routing Protocol	
Mobility/Traffic Level	*Feature*	*AODV*	*TRBPF*
Low/low	Packet delivery ratio	High	High
	Latency	Low	Medium
	Routing overhead	Low	Medium
High/high	Packet delivery ratio	High	High
	Latency	Medium	Medium
	Routing overhead	High	Medium

the end-to-end delay (latency) of packets, and the total number of control packets to data packets (routing overhead). Table 7.8 provides a general comparison between AODV and TBRPF for low- and high-mobility scenarios, with a low traffic rate occurring under low mobility, while a high traffic rate is presumed to occur under high mobility. Note that the routing overhead is highly dependent upon the mobility mode of operation. Because AODV transmits many small routing control packets, it is well suited for both low- and high-mobility operations. Similarly, because TBRPF is proactive and has a constant routing overhead, it is also very suitable for both low- and high-mobility operations.

The latency of AODV for low-mobility and low-traffic operations is slightly better than that of TBRPF, because AODV uses small routing control packets. However, for high-mobility and high-traffic operations latency associated with either protocol becomes very similar. The last comparison shown in Table 7.8 concerns routing overhead. Although AODV uses smaller routing control packets than TBRPF, the former is only more efficient when supporting low-traffic environments. As traffic increases, TBRPF, which represents a proactive protocol, becomes more efficient as it has a constant routing overhead.

7.2 The Intelligent Roadway

In this concluding section we will examine how intelligence can be provided to vehicles that use highways, roads, and streets about conditions they can expect to encounter. Because ad hoc networks can be used to extend the range of information dissemination, we will also discuss the relationship between the two technologies.

7.2.1 Roadway Design

Through the use of wireless access points or even simple broadcast stations located at strategic locations, such as exit ramps or at the beginning of a curved incline or

Table 7.9 Locations on Roadways that Could Communicate with Vehicles

Lane markers
Lane direction
Road junction
Traffic light
Road exit
Temporary obstacles

steep hill, information can be presented to approaching vehicles. Table 7.9 lists some of the locations on a roadway where information could be broadcast either directly to an approaching vehicle or via a Vehicle Mobile Ad Hoc Network (VANET).

7.2.1.1 Lane Markers

Lane markers not only identify where a vehicle should travel, but in addition represent a safety feature, as the marker prevents vehicle drift that can result in a collision. Unfortunately, drivers who are tired, inebriated, or simply not paying attention to the road may drift into the wrong lane, causing an accident.

Although lane markers prevent accidents, they are passive devices. If some lane markers are supplemented with wireless transmitters, they could be used to communicate with a vehicle operator, resulting in an audio or visual warning when lane drift occurs that can be expected to enhance road safety.

7.2.2 Lane Direction

As you travel on different highways you will periodically note signs with the message "Do Not Enter" placed at locations where it is relatively easy for an inexperienced or confused vehicle operator to drive onto a lane where vehicles are traveling in the opposite direction. Although such signs have significantly reduced head-on collisions, they are difficult to view at night and during inclement weather. Thus, they represent another location where the broadcast of a warning message would be appropriate. In fact, using a GPS receiver and applicable software, the warning message could be ignored unless the vehicle operator incorrectly positions his or her vehicle so that it might enter the restricted lane. Thus, it is possible that many roadway information broadcasts could be ignored by vehicles and never turned into an alarm that obtains the attention of the vehicle's operator.

7.2.3   Road Junction

The location where two or more roads intersect represents a road junction that is statistically prone to accidents even when both warning and stop signs are used to prevent accidents. Thus, a road junction would represent another location from which a warning broadcast would be appropriate. That broadcast in the future could be integrated into the electronics of a vehicle and automatically slow the vehicle as it approaches the intersection.

7.2.4   Traffic Light

At one time traffic lights were completely passive devices, simply turning on and off a sequence of colors from green to yellow and then red. Today many traffic lights in urban areas are integrated with video cameras placed at strategic locations to take photographs of drivers and license plates of vehicles that drive through red lights. Thus, after running a red light a motorist might receive via mail a traffic fine that shows the vehicle operator running a red light, as well as a form to complete that is returned with a check to pay the fine.

In addition to integrating video cameras with traffic lights, it is possible to broadcast the state of the traffic light. With appropriate software, a vehicle could receive the traffic signal and adjust its speed automatically to either come to a gradual stop as the light changes from yellow to red or slightly accelerate to ensure the vehicle passes the intersection while the light is green. Thus, in addition to providing a safety measure, the broadcast of the state of a traffic light could be used to regulate traffic flow to enhance a vehicle's mileage.

7.2.5   Road Exit

As traffic exits a highway, vehicles can encounter a wide variety of situations. These range from stop signs and traffic lights to obstructions caused by construction and various road signs indicating the direction and mileage to gas stations, restaurants, and rest stops. Through broadcasted information about the traffic exit, the operators of approaching vehicles can be warned about existing construction and informed about the availability of nearby gas and service stations as well as dining facilities. Thus, the use of vehicle-to-vehicle transmission upon receipt of a broadcast to an approaching vehicle could extend the range of information dissemination and provide vehicle operators with additional time to consider taking an exit ramp.

7.2.6   Temporary Obstacles

As vehicles traverse different types of roads, they will typically encounter lanes that are being resurfaced, roads where potholes are being filled, and other obstructions that may require the vehicle to merge into a different lane as well as lower its speed. By placing a broadcast transmitter at such locations, oncoming traffic could be warned about the need to slow down and perhaps merge into a different lane. Thus, temporary obstacles represent another traffic condition where transmission of broadcast information could supplement conventional traffic signs.

7.2.7   Transmission Methods

There are two basic methods by which information can be transmitted to approaching vehicles: infrastructure to vehicle and vehicle to vehicle. When a Vehicle Mobile Ad Hoc Network is formed, it also becomes possible for the initial infrastructure-to-vehicle transmission to be relayed through the formed VANET.

7.2.7.1   Infrastructure to Vehicle

Because initially only a small number of vehicles and roadways will be equipped to transmit and receive data, this author believes that at first the infrastructure-to-vehicle transmission model will represent the primary method used for roadway communications. Figure 7.16 illustrates an example of this model.

In examining Figure 7.16, note that of the five vehicles shown on the highway, only one is within range of a roadside transmitter. In addition, although that vehicle is within range of the transmitter, it may or may not have the capability to receive data broadcast.

Because there are over 150 million vehicles on the road in the United States, while production averages approximately 16 million vehicles per year, it will probably take at least a decade and perhaps longer until a majority of vehicles on the

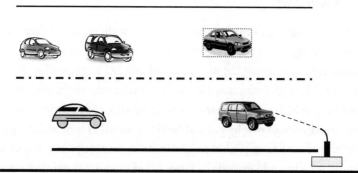

Figure 7.16   Infrastructure-to-vehicle transmission.

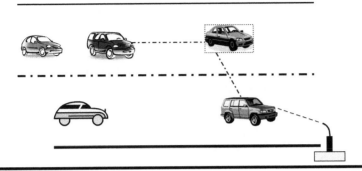

Figure 7.17   Extending roadway transmission distance via a vehicle-to-vehicle network.

road are capable of receiving data from roadside transmitters. Because it may take even longer to have a majority of vehicles on the road capable of forming a Vehicle Ad Hoc Network, this author believes that the intelligent roadway will initially be focused on providing information directly to vehicles. Then, as more vehicles become equipped with a VANET capability, the range of roadway transmission will be extended via mobile mesh networks.

7.2.7.2   Vehicle to Vehicle

As previously mentioned, the vehicle-to-vehicle transmission method can be viewed as a mechanism to increase the range of a roadside transmission. This situation is illustrated in Figure 7.17. In this example, three of the five vehicles on the roadway have a vehicle-to-vehicle transmission capability and are within transmission distance of one another. Thus, the vehicle nearest to the roadside transmitter receives the broadcast of information and relays that data to other vehicles in the Vehicle Ad Hoc Mobile Network.

In the example shown in Figure 7.17, three vehicles are close enough together to form a mesh network. Thus, the transmission from the roadway can be extended, allowing vehicles with a VANET capability to have more time to consider the information contained in the transmission.

7.2.8   The Evolving Smart Vehicle

In concluding this chapter, this author will examine how the future smart vehicle may operate. To do so, let us discuss the communications electronics that will eventually be contained in the evolving smart vehicle. Table 7.10 lists the major communications-related components of the evolving smart vehicle. Note that the display can be either part of the center console navigation system or an independent

Table 7.10 The Communications Components of the Emerging Smart Vehicle

Microprocessor
Wireless networking transmitter/receiver
Forward-looking radar
Side radar GPS
Cellular transmitter/receiver
Event data recorder

heads-up military aircraft type of display, which has been an option on certain GM vehicles for several years.

In the following paragraphs we will discuss each of the communications components of the emerging smart vehicle.

7.2.8.1   Microprocessor

The microprocessor receives input from the various communications devices. Such input is used to display warnings and other information on a display, via an audio alarm, and perhaps eventually via a voice synthesis system.

Another function of the microprocessor will be to operate any routing protocol selected for the formation of a Vehicle Ad Hoc Mobile Network. Thus, vehicle manufacturers could elect to use one high-performance processor or subdivide work by using two or more microprocessors.

7.2.8.2   Wireless Networking Transmitter/Receiver

The wireless network transmitter/receiver enables the vehicle to receive roadway broadcasts as well as relay such broadcasts if the vehicle is within range of other VANET-compatible vehicles. There are significant benefits that can occur by integrating the VANET system with a GPS, which is why the two will more than likely be interconnected in the emerging smart vehicle.

7.2.8.3   Forward Radar

Forward radar will be used to detect any forward obstacles at distances up to 400 to 500 ft, depending upon the contour of the terrain. This radar may operate in the unlicensed wideband (UWB) frequency range in the 60-GHz spectrum and will more than likely be integrated into a vehicle's braking system. Thus, the sudden appearance of an obstacle in front of a vehicle or the rapid braking of a leading vehicle would result in the forward radar immediately detecting a closing of the distance between vehicles, which would result in either an automatic adjustment to the vehicle's cruise control, if enabled, or initiation of the vehicle's braking.

7.2.8.4   Side Radar

The purpose of side radar is to detect the presence of vehicles in a vehicle's blind spots. Because the primary purpose of side radar is to look for obstacles in the blind spots of a vehicle, it will operate on low power and have a transmission range of less than 10 ft.

As an alternative to the use of side radar, it is possible that future vehicles will be equipped with miniature TV cameras or use ultrasonic sensors. At the present time, vehicle manufacturers are testing all three technologies as a mechanism to reduce accidents caused by lane changes.

7.2.8.5   GPS

GPS will be used to provide a vehicle's location. Through the microprocessor GPS will be integrated to the navigation system as well as crash sensors, enabling the deployment of an air bag to be transmitted to a monitoring station.

7.2.8.6   Cellular Transmission

The cellular transmitter can be viewed along with the GPS subsystem as part of a vehicle's telemetric system. Here the deployment of an air bag would automatically result in a cellular call to a monitoring location, to include the GPS position of the vehicle. In addition, by pressing a button on the console, the vehicle operator can use a concierge service to request the location of a vehicle repair shop or ask another query without having to enter information into a navigation system while driving.

7.2.8.7   Event Data Recorder

Similar to an aircraft's black box, the event data recorder is used in vehicles to store all important parameters, such as acceleration, position, velocity, and status of subsystems, to include tire pressure. By examining the contents of the event data recorder, it may be possible to determine the cause of vehicle problems ranging from a collision to the failure of an engine to start.

7.2.9   Summary

As indicated in this section, the use of applicable transmitters along roadways, along with the emerging communications capability of the smart vehicle, can be expected to provide a significant enhancement to both road safety and driver and passenger requests for information. Although we are many years removed from the true smart vehicle, and at the present time do not know the exact components of the vehicle of the future, we can be assured that its benefits will result in its eventual deployment.

Index

Milton Keynes UK
Ingram Content Group UK Ltd.
UKHW040056071024
449327UK00019B/597